Raja Bahar Khan Soomro

Environmental Education & Lifelong Learning for Sustainable Futures

Raja Bahar Khan Soomro

Environmental Education & Lifelong Learning for Sustainable Futures

A Critical Analysis

Noor Publishing

Imprint
Any brand names and product names mentioned in this book are subject to trademark, brand or patent protection and are trademarks or registered trademarks of their respective holders. The use of brand names, product names, common names, trade names, product descriptions etc. even without a particular marking in this work is in no way to be construed to mean that such names may be regarded as unrestricted in respect of trademark and brand protection legislation and could thus be used by anyone.

Cover image: www.ingimage.com

Publisher:
Noor Publishing
is a trademark of
Dodo Books Indian Ocean Ltd. and OmniScriptum S.R.L publishing group

120 High Road, East Finchley, London, N2 9ED, United Kingdom
Str. Armeneasca 28/1, office 1, Chisinau MD-2012, Republic of Moldova, Europe
Printed at: see last page
ISBN: 978-620-7-47902-3

Environmental Education & Lifelong Learning for Sustainable Futures: A Critical Analysis

Rationale

The Earth is our home, unique in the vastness of space, where life thrives in a delicate balance nurtured over thousands of years. But in recent centuries, as industries grew and technology advanced, humans have upset this balance. Driven by a quest for wealth and comfort, we have harmed the environment, disrupting its natural harmony and vitality. As a result of all this, we are now facing serious consequences like climate change, pollution, and threats to our own survival.

The book *"Environmental Education & Lifelong Learning for Sustainable Futures: A Critical Analysis"* stresses the urgent need to protect our environment through knowledge and action. It aims to share essential scientific insights about how ecosystems work including the impact of human actions. It also argues that we are at a critical moment where we must work together to reduce harm to nature and find sustainable ways forward for our future generations.

A key idea in the book is that environmental education can transform lives. It proposes an approach to learning that goes beyond classrooms, teaching people the skills and values needed to tackle environmental problems. The book also emphasizes the importance of everyone understanding how ecosystems function from schools to governments to local communities. By fostering a deep respect for nature and its value, this book aims to inspire us all to adopt sustainable practices and care responsibly for natural resources.

Furthermore, the book examines the main causes of environmental damage, such as unchecked industry, deforestation, rising carbon emissions, irresponsible behaviour towards environment preservation, and consumerism. It also highlights that how these activities affect vulnerable communities and future generations the most, creating environmental injustices and disruptions. By outlining these interconnected challenges, the book calls for global awareness and action to promote fair and sustainable development.

Overall, *"Environmental Education & Lifelong Learning for Sustainable Futures: A Critical Analysis"* urges its readers to take responsibility with their capacity and protect our planet's legacy. It advocates a shift towards sustainable practices that respect nature, promote fairness, and strengthen our economies. By deepening our understanding of environmental issues and making ethical choices, the book aims to empower individuals and communities to preserve biodiversity and ensure a healthy future for our coming generations.

In conclusion, the book argues that lifelong environmental education is crucial for adapting to the environmental challenges of the 21st century. It stresses the need to balance human progress with ecological health, paving the way for sustainable futures that honour our natural world and benefit future generations.

Author
Raja Bahar Khan Soomro

Foreword

In the grand tapestry of time, Earth stands as a marvel of creation, a singular jewel in the cosmos, where life thrives in a delicate equilibrium sculpted by millennia of natural processes. However, the relentless march of human progress over the past few centuries has cast a long shadow over this pristine balance. Driven by an insatiable pursuit of wealth and comfort, we have engaged in practices that have profoundly disrupted the natural harmony of our world, leading to an array of severe environmental challenges. Climate change, pollution, and the erosion of biodiversity are now among the stark realities we face, presenting urgent threats to our very survival.

In this critical juncture of human history, the call for change has never been more pressing. "Environmental Education & Lifelong Learning for Sustainable Futures: A Critical Analysis," authored by Raja Bahar Khan Soomro, emerges as a beacon of hope and enlightenment in this time of environmental peril. This book is not merely an academic exploration but a passionate plea for action, a call to recognize the indispensable role of environmental education in shaping a sustainable future.

The core of author's argument revolves around the transformative power of environmental education. This book makes a compelling case for an approach to learning that transcends traditional classroom boundaries, advocating for a model that integrates knowledge with practical action. It underscores the necessity of fostering a profound understanding of how ecosystems function, and the impacts of human actions on these systems. By embedding these principles in educational frameworks, from schools to community organizations and beyond, Soomro envisions a society equipped with the skills and values essential for addressing and mitigating environmental challenges.

Central to this book is the notion that education can be a catalyst for profound change. The text delves into the systemic causes of environmental degradation, unrestrained industrial activities, deforestation, rising carbon emissions, and unsustainable consumption patterns. It thoughtfully addresses how these issues disproportionately affect vulnerable communities, perpetuating cycles of environmental injustice and instability. Soomro's analysis brings to light the urgent need for equitable and sustainable development strategies that not only address environmental concerns but also promote social fairness.

"Environmental Education & Lifelong Learning for Sustainable Futures" is both a diagnosis and a prescription. It critically examines the multifaceted causes of environmental damage while advocating for a paradigm shift towards sustainable practices. The book's call to action is both ambitious and necessary: to engage in lifelong learning that aligns human progress with ecological stewardship. By adopting this approach, we can aspire to create a future where economic growth does not come at the expense of environmental health, but rather in harmony with it.

As we stand on the threshold of an uncertain future, Soomro's work offers invaluable insights and practical guidance. It challenges us to rethink our relationship with nature, urging a collective commitment to sustainable living that honours the legacy of our planet and secures a healthy environment for future generations. This book is a vital resource for educators, policymakers, and concerned citizens alike, serving as a clarion call to action in our quest to safeguard the delicate balance of our world.

It is with great anticipation and hope that I commend "Environmental Education & Lifelong Learning for Sustainable Futures: A Critical Analysis" to all readers. Let this be a guiding light in our efforts to forge a sustainable path forward, embracing the wisdom of environmental education to foster a world where humanity and nature thrive together.

Abdul Basit Soomro

M.Phil Edn & PhD Edn Scholar

Acknowledgements

The journey of writing "Environmental Education & Lifelong Learning for Sustainable Futures: A Critical Analysis" has been a profound and enlightening experience, made possible through the support, inspiration, and collaboration of many individuals and institutions. It is with deep gratitude that I acknowledge those who have contributed to this work.

First and foremost, I extend my heartfelt thanks to the scholars and researchers whose work has been foundational to this book. Their invaluable insights and ground-breaking research have provided the framework upon which this analysis is built. To the authors and contributors referenced throughout the text, your pioneering studies and passionate advocacy for environmental issues have greatly enriched my understanding and shaped the arguments presented herein.

I am profoundly grateful to the institutions and organizations that have supported and guided this endeavour. Special thanks to the environmental think tanks, educational institutions, and non-governmental organizations whose resources and expertise have been instrumental in my research. Your commitment to environmental sustainability and education has been a source of inspiration and guidance.

I also offer my sincere thanks to my colleagues and mentors who have provided constructive feedback and encouragement throughout the writing process. Your thoughtful critiques and supportive words have been essential in refining the content and ensuring the accuracy of the analysis.

To the dedicated educators and practitioners in the field of environmental education, your work in promoting and implementing sustainable practices has been a beacon of hope. Your real-world applications and innovative teaching methods have illustrated the transformative potential of environmental education, a theme central to this book.

I also wish to express my appreciation to the reviewers and editors who have meticulously reviewed the manuscript. Your keen eyes and detailed comments have greatly improved the clarity and quality of this work.

Lastly, to my family and friends, your unwavering support and understandings have been a cornerstone throughout this journey. Your encouragement has been a source of strength, allowing me to navigate the challenges and triumphs of this writing process.

This book is a reflection of collective efforts and shared knowledge. It stands as a testament to the collaborative spirit that is essential in addressing the environmental challenges we face. I hope that the insights and recommendations presented herein contribute meaningfully to the on-going dialogue about environmental education and sustainability, and inspire action towards a more balanced and equitable future for our planet.

Thank you all for your contributions and support.

Raja Bahar Khan Soomro

Index

Chapter 1: Introduction

This chapter explores the critical role of environmental education in addressing the profound impacts of human activities on our planet. It contrasts our destructive actions, such as pollution, deforestation, and industrialization, with our investment in space exploration, highlighting the paradox of advancing technology while neglecting Earth's well-being. The chapter underscores that scientific and technological progress, while beneficial, has led to significant environmental degradation, including global warming, biodiversity loss, and severe weather events.

It emphasizes the urgent need for environmental education as a tool to foster awareness and promote sustainable practices. Environmental education extends beyond academic learning to instil a deep sense of responsibility and stewardship across all sectors of society. By integrating environmental principles into lifelong learning, the chapter argues that individuals and communities can better adapt to and address environmental challenges.

Proactive measures, such as stringent regulations, renewable energy promotion, and conservation efforts, are essential for mitigating further damage. The chapter advocates for a collective effort, including educational institutions, to build a culture of environmental awareness and action, ensuring a sustainable future for generations to come.

1.1 Importance of Environmental Education: An Overview

It's an unfortunate that on the one hand we are damaging and destroying our beautiful planet earth due to our own actions, and on the other hand we are spending billions of dollars in space exploration and searching for the possibilities of life on other planets. It is a bitter fact that today our planet's biodiversity and natural richness are in danger due to our routine actions. In the name of advancement and modernization we have actually crossed the limits as far as environmental hazards are concerned. However, the good news is this that the world has now started realising that our own actions are causing great threats to our future generations. Similarly, the world has now started realising that we have disturbed the natural balance of our planet and as a result of it we are experiencing severe consequences in terms of global warming and other environmental hazards.

Basically, our environment encompasses both biotic (living) and abiotic (non-living) components, where natural and synthetic materials coexist. The scientific advancements of the 20th century have profoundly impacted our natural surroundings, ushering in an era of unprecedented technological progress. While these developments have catapulted us into an advanced world, they have also exacted a significant toll on our environment. The rapid population growth in the latter half of the 20th century, coupled with densifying societies, has precipitated an insatiable rise in human consumption and demand.

Consequently, this surge has not only expanded economic markets and fostered wealth creation but has also intensified environmental pressures.

Scientific innovations facilitated the mass production of synthetic goods to meet burgeoning human needs. Concurrently, industrialization and the discovery of underground oil reservoirs have spurred unbridled carbon emissions and energy consumption. Regrettably, the implications of such extensive emissions were underestimated, leading to severe environmental repercussions. Additionally, the development and deployment of powerful explosives and quest for mass-destructive weaponry have inflicted irreparable harm on our natural landscapes.

The consequences of human activities extend beyond terrestrial degradation to include deforestation, landfills overflowing with waste, and the destabilization of mountainous regions, all of which disrupt the delicate equilibrium of the Earth's systems. Heightened sea levels, attributable to these activities, have precipitated calamitous phenomena like hurricanes and tsunamis, underscoring the direct link between human actions and environmental instability.

Moreover, the proliferation of pollutants such as airborne particulates, contaminated soils, tainted water bodies, incessant noise, and marine debris has fostered the emergence and propagation of various debilitating diseases. These pollutants pose a grave threat to ecosystems, imperilling diverse species that form the bedrock of biodiversity.

Presently, the world grapples with formidable challenges posed by climate change and global warming. These issues have reached critical thresholds, necessitating immediate short-term and sustained long-term responses. Addressing these challenges demands concerted efforts from all segments of society, including political entities, international organizations, grassroots activists, civil societies, and educational institutions. Their collective engagement is essential to formulating and implementing effective strategies aimed at achieving a sustainable environment.

In response to these pressing concerns, proactive measures must be swiftly enacted to mitigate further environmental degradation. This includes stringent regulations to curb carbon emissions, promote renewable energy sources, prioritizing recycling synthetic products, and foster eco-friendly industrial practices. Additionally, safeguarding natural habitats through comprehensive conservation initiatives and reforestation efforts can certainly help restore ecological balance.

Furthermore, education and awareness campaigns play a pivotal role in fostering a culture of environmental stewardship and promoting responsible consumption patterns. By instilling ecological values and principles in future generations, educational institutions can cultivate a cadre of environmentally conscious global citizens capable of effecting positive change.

Overall, while humanity has reaped immense benefits from scientific progress and industrialization, the unintended consequences on our environment underscore an urgent need for concerted action. By embracing sustainable practices and fostering global cooperation, we can forge a path towards a resilient and harmonious coexistence with our natural surroundings. The time to act decisively is now, lest we face irreparable harm to our planet and future generations.

1.2 Definition of Key Terms & Concepts

The critical discourse analysis of this book is mainly based on three key terms including intersection and synergies among them. These three key terms include environmental education, lifelong learning, and sustainable futures. Before moving forward, it is crucial to define all these terms in its true context so that you may grasp the main idea and get useful insights from this book. Let us now define these key terms one by one.

1.2.1 Environmental Education: Environmental education is a multifaceted concept that encompasses a wide range of topics and disciplines related to the preservation and sustainability of our environment. This broad field goes beyond merely integrating environmental studies into formal education systems. It extends to fostering an eco-conscious mind-set and promoting sustainable practices among entrepreneurs, designers, scientists, biologists, chemists, engineers, energy experts, politicians, and other professionals.

Environmental education aims to instil a deep sense of environmental responsibility and stewardship at both individual and community levels. This approach encourages everyone, from individuals to entire nations, to adopt and advocate for eco-friendly practices and policies. By embedding environmental education into the fabric of various professional fields and everyday life, we cultivate a society that is better equipped to tackle environmental challenges and pursue sustainable development.

One of the key aspects of environmental education is its integration with lifelong learning. Lifelong learning is the continuous pursuit of knowledge and skills throughout an individual's life. When environmental education is intertwined with lifelong learning, it ensures that people remain informed about environmental issues and advances in sustainability practices, regardless of their age or professional background.

For example, an entrepreneur informed by environmental education might innovate sustainable business practices, while a politician might advocate for policies that reduce carbon emissions. Similarly, engineers and scientists can develop technologies that mitigate environmental impact, and educators can incorporate sustainability into their teaching methods.

In essence, environmental education is not a one-time effort but an on-going process that evolves with societal and technological advancements. It requires the active participation of all sectors of society to foster a culture of environmental stewardship. By linking environmental education with lifelong learning, we can create a more informed, proactive, and sustainable world, capable of addressing current and future environmental challenges effectively.

1.2.2 Lifelong Learning: Lifelong learning is the continuous process of acquiring new knowledge, honing professional skills, and adapting to changes. This on-going journey of education and self-improvement is essential in an ever-evolving society where demands and technologies are constantly shifting. The concept of lifelong learning encompasses more than just formal education; it includes informal learning, self-study, and experiential learning throughout our lives.

In today's fast-paced world, the importance of lifelong learning cannot be overstated. The rapid advancement of technology and the globalization of economies mean that the skills and knowledge required to succeed are continually changing. To remain competitive and effective in our professional and personal lives, we must commit to constantly updating our skill sets and staying informed about new developments in our fields.

Lifelong learning is not just about professional development. It also enriches our personal lives, broadens our perspectives, and enhances our understanding of the world. Engaging in lifelong learning helps us to stay mentally active, curious, and engaged, fostering a sense of fulfilment and purpose.

Moreover, lifelong learning fosters adaptability and resilience. By continually learning and adapting, we become better equipped to handle unexpected challenges and changes, whether in our careers or personal lives. This adaptability is crucial in a world where the only constant is change.

The process of lifelong learning is inherently self-motivated and self-directed. It requires a proactive attitude and a willingness to seek out new experiences and knowledge. This might involve enrolling in courses, attending workshops, reading extensively, or even traveling to gain new perspectives.

In conclusion, lifelong learning is a vital and on-going process that plays a crucial role in personal and professional development. It helps us to stay relevant, adaptable, and intellectually engaged throughout our lives, enabling us to meet the demands of an ever-changing society with confidence and competence.

1.2.3 Sustainable Futures: When we talk about sustainable futures, we envision an environment that is eco-friendly, socio-economically balanced, and conducive to human well-being for the generations to come. This concept extends beyond just human interests, embracing a holistic perspective that includes the well-being of all species and living organisms. Every species, whether currently thriving or at risk of extinction, plays a crucial role in maintaining the natural balance of our planet's rich biodiversity.

Creating a sustainable future involves implementing practices that protect and nurture the environment. This means reducing pollution, conserving natural resources, and transitioning to renewable energy sources. It also involves fostering economic systems that are both resilient and equitable, ensuring that all communities have access to the resources they need to thrive. Sustainable economic practices should promote fair trade, support local economies, and minimize environmental impact.

Social sustainability is equally important. It requires building societies that are inclusive, just, and supportive of human rights. This involves creating educational opportunities, ensuring access to healthcare, and promoting social justice. By addressing issues like poverty, inequality, and social exclusion, we can build communities that are stronger and more resilient.

However, the concept of sustainable futures is not limited to human concerns. It acknowledges the interdependence of all life forms on Earth. Preserving the habitats and ecosystems that support diverse species is vital. Each organism, from the smallest insect to the largest mammal, contributes to the ecological balance and overall health of the planet. Protecting endangered species and restoring degraded ecosystems are critical components of a sustainable future.

In essence, sustainable futures require a comprehensive approach that integrates environmental stewardship, economic stability, and social equity. By working towards these goals, we can create a world where both people and nature can coexist and flourish, ensuring the health and prosperity of our planet for future generations.

1.3 Objectives of the Book & Outline of Chapters

The primary objectives of this book are multifaceted, aiming to instil a deep sense of awareness among readers about the critical importance of environmental education. The author aspires to make a meaningful contribution to society, ensuring that future generations inherit a healthy, sustainable, and eco-friendly environment. To achieve these goals, the book has been meticulously structured, with each chapter thoughtfully chosen to provide comprehensive insights into the environment, lifelong learning, and sustainable futures.

One of the key themes of the book is lifelong learning for environmental literacy, emphasizing the continuous acquisition of knowledge and skills necessary to understand and address environmental issues. This concept is vital as it empowers individuals to make informed decisions and take responsible actions to protect the environment.

The book delves into sustainability concepts and practices, offering readers practical approaches to incorporate sustainability into their daily lives and communities. It explores the foundational principles of environmental education, highlighting its role in fostering an informed and engaged citizenry capable of driving positive environmental change.

Innovative scientific fields such as synthetic biology and microchemistry are also examined, focusing on their potential to address environmental challenges. Synthetic biology's environmental implications and microchemistry's innovative solutions are discussed in detail, showcasing their roles in creating sustainable technologies.

Energy, a critical component of environmental discussions, is addressed with an exploration of its opportunities and challenges. The book discusses the intersection of physics and environmental science, illustrating how fundamental scientific principles can be applied to understand and solve environmental problems.

Additionally, the concept of the circular economy is presented, emphasizing the importance of reducing waste and promoting the reuse and recycling of materials. The book also reviews environment-related policies, providing readers with an understanding of the regulatory frameworks that shape environmental action and governance.

In summary, this book is a comprehensive guide designed to equip readers with the essential knowledge and tools to contribute to a sustainable future. Through its carefully selected chapters, it offers a holistic approach to environmental education, fostering a culture of awareness and responsibility towards our planet.

1.4 References

1. *Carson, R. (1962). Silent Spring. Houghton Mifflin.*
2. *Giddens, A. (2009). The Politics of Climate Change. Polity Press.*
3. *Harrison, R. (2010). Environmental Education and Communication: An Integrated Approach. Routledge.*
4. *Hawken, P. (1999). Natural Capitalism: Creating the Next Industrial Revolution. Little, Brown and Company.*
5. *IPCC (2021). Climate Change 2021: The Physical Science Basis. Cambridge University Press.*
6. *Kuhlman, T., & Farrington, J. (2010). What is Sustainability?. Sustainability, 2(11), 3436-3448.*
7. *Meadows, D. H., Meadows, D. L., Randers, J., & Behrens III, W. W. (1972). The Limits to Growth. Universe Books.*
8. *Miller, G. T., & Spoolman, S. (2012). Environmental Science: Sustaining Your World. Cengage Learning.*
9. *UNESCO (1975). The Tbilisi Declaration. UNESCO.*
10. *World Commission on Environment and Development (1987). Our Common Future. Oxford University Press.*

Chapter 2: Lifelong Learning for Environmental Literacy

Lifelong learning is pivotal in nurturing environmental awareness and stewardship, essential for tackling the planet's pressing challenges such as climate change and biodiversity loss. This on-going educational process empowers individuals to stay informed about evolving environmental issues and adapt their behaviours accordingly. By continuously acquiring knowledge and skills, individuals can bridge the gap between formal education and real-world applications, promoting sustainable practices and responsible decision-making.

Engaging in lifelong learning fosters a deeper understanding of the interconnectedness between human activities and natural systems, motivating individuals to adopt more sustainable practices. For example, awareness of one's carbon footprint can lead to more conscious choices like reducing energy consumption and minimizing waste. Furthermore, lifelong learning enhances critical thinking and problem-solving skills, which are crucial for addressing complex environmental issues and encouraging interdisciplinary solutions.

In the workplace, lifelong learning supports the implementation of sustainable business practices, contributing to broader sustainability goals while enhancing organizational competitiveness. Additionally, it enables informed civic engagement, empowering individuals to participate in environmental policy discussions and advocacy. Overall, by integrating continuous education into all stages of life, societies can cultivate a culture of sustainability and ensure the health and well-being of our planet for future generations.

2.1 Importance of Lifelong Learning in Fostering Environmental Awareness & Stewardship: Lifelong learning plays a crucial role in fostering environmental awareness and stewardship, essential for addressing the complex and evolving challenges facing our planet. This continuous process of acquiring knowledge, skills, and attitudes enables individuals to remain informed and engaged with environmental issues throughout their lives, promoting sustainable behaviours and responsible decision-making.

The dynamic nature of environmental challenges, such as climate change, biodiversity loss, and pollution, necessitates an on-going educational approach. As scientific understanding and technological advancements evolve, lifelong learning ensures that individuals can adapt and respond effectively. It bridges the gap between formal education and real-world application, allowing people to continuously update their knowledge and skills to address emerging environmental issues.

Engaging in lifelong learning fosters a deeper understanding of thc interconnectedness of natural systems and human activities. It helps individuals recognize the impact of their actions on the environment and motivates them to adopt sustainable practices. For instance, learning about the carbon footprint associated with daily activities can lead to more conscious choices, such as reducing energy consumption, minimizing waste, and supporting sustainable products and services.

Lifelong learning also empowers individuals to participate actively in environmental stewardship. It cultivates a sense of responsibility and agency, encouraging people to take part in community initiatives, conservation efforts, and policy advocacy. For example, community-based environmental education programs provide opportunities for individuals to engage in local conservation projects, enhancing their understanding of ecological processes and fostering a collective sense of stewardship.

Furthermore, lifelong learning promotes critical thinking and problem-solving skills, essential for addressing environmental challenges. It encourages individuals to question existing practices, seek innovative solutions, and collaborate across disciplines. This interdisciplinary approach is vital for tackling complex environmental issues that require integrated strategies and diverse perspectives.

In the workplace, lifelong learning equips professionals with the knowledge and skills needed to implement sustainable practices within their industries. For example, on-going training in green technologies and sustainable business practices can help companies reduce their environmental impact and contribute to broader sustainability goals. This not only benefits the environment but also enhances the long-term viability and competitiveness of businesses.

Moreover, lifelong learning supports the development of environmental policies and governance. Educated and informed citizens are better equipped to engage in policy discussions, advocate for effective regulations, and hold decision-makers accountable. This civic engagement is crucial for driving systemic changes that promote environmental sustainability at local, national, and global levels.

Educational institutions, governments, and organizations play a vital role in promoting lifelong learning for environmental awareness and stewardship. They can provide accessible resources, create supportive learning environments, and foster a culture of continuous education. Initiatives such as online courses, workshops, public lectures, and community programs make environmental education more accessible and inclusive, reaching diverse audiences and enabling widespread participation.

In conclusion, lifelong learning is essential for fostering environmental awareness and stewardship. It equips individuals with the knowledge, skills, and attitudes needed to address on-going and emerging environmental challenges, promotes sustainable behaviours, and empowers active participation in environmental conservation and policy advocacy. By embracing lifelong learning, society can cultivate a culture of sustainability that ensures the health and well-being of our planet for future generations.

2.2 Educational Strategies & Programs Promoting Environmental Literacy at Different Stages of Life: Educational strategies and programs promoting environmental literacy are crucial at every stage of life, as they help individuals understand and address the complexities of environmental issues. Early childhood education often introduces basic concepts of nature and the environment through interactive and engaging activities. Young children learn about the natural world through stories, songs, and hands-on experiences such as planting gardens or exploring local parks. These early experiences lay the foundation for a lifelong appreciation and understanding of the environment.

As children progress into primary and secondary education, the curriculum typically becomes more structured and detailed. Environmental science is often integrated into subjects like biology, chemistry, and geography. Programs such as school gardens, recycling initiatives, and field trips to nature reserves or science museums help to reinforce classroom learning with real-world experiences. For example, the Eco-Schools program operates in many countries, providing a framework for schools to become more sustainable through student-led projects and activities.

At the tertiary education level, universities and colleges offer specialized courses and degrees in environmental science, sustainability, and related fields. These programs provide in-depth knowledge and research opportunities, preparing students for careers in environmental protection, policy, and innovation. Institutions often engage students in community-based projects, internships, and partnerships with environmental organizations, which enhance practical skills and professional networks.

For adults, environmental education continues through workplace training, community programs, and online learning platforms. Companies increasingly recognize the importance of environmental literacy for their employees, implementing sustainability training and green practices in the workplace. Community education programs, such as those offered by local governments, non-profits, and environmental organizations, provide workshops, seminars, and volunteer opportunities focused on issues like energy conservation, waste reduction, and biodiversity.

In recent years, digital technology has expanded access to environmental education. Online courses, webinars, and social media campaigns reach a broad audience, making it easier for people of all ages to stay informed about environmental issues and solutions. Platforms like *Coursera, TEDx,* and *edX* offer online courses and talk sessions on sustainability and environmental management, often in partnership with leading universities, research scholars, and experts.

Moreover, lifelong learning initiatives encourage continuous engagement with environmental topics. Public libraries, museums, and nature centres often host lectures, exhibitions, and educational programs for adults and families. These institutions play a crucial role in providing accessible and on-going opportunities for environmental education. Lifelong learning initiatives can be extended at different market places, public gatherings, sporting events, and religious ceremonies.

Overall, promoting environmental literacy across different life stages requires a multifaceted approach that combines formal education, community involvement, as well as digital resources. By engaging individuals at every stage of life, these strategies help build a society that is knowledgeable, proactive, and committed to sustaining the environment for future generations.

2.3 Role of Educators, Community Leaders, & NGOs in Promoting Lifelong Environmental Learning: Educators, community leaders, and non-governmental organizations (NGOs) play pivotal roles in promoting lifelong environmental learning, each contributing uniquely to raising awareness and fostering sustainable practices within society.

Educators are at the forefront of imparting environmental knowledge and instilling values of sustainability. They introduce students to critical concepts of ecology, conservation, and sustainable development from an early age, embedding these principles into the curriculum across various subjects. By integrating real-world environmental issues into lessons, educators help students understand the relevance of these topics to their lives and future. Additionally, hands-on activities such as field trips, experiments, and community projects encourage active learning and foster a deeper connection to nature. Educators also serve as role models, demonstrating sustainable practices in their personal and professional lives, thereby inspiring students to adopt similar behaviours.

Community leaders hold significant influence and can mobilize local populations towards environmental stewardship. They can advocate for sustainable practices within their communities, such as promoting recycling programs, organizing clean-up drives, and encouraging the use of renewable energy sources. By leveraging their positions of authority and trust, community leaders can facilitate discussions on environmental issues, creating platforms for residents to voice concerns and contribute ideas for local sustainability initiatives. Furthermore, they can collaborate with local schools and organizations to support environmental education programs, ensuring that lifelong learning opportunities are accessible to all community members.

NGOs play a crucial role by bridging the gap between education, advocacy, and action. They often provide resources and expertise that may be beyond the reach of formal educational institutions and local governments. NGOs organize workshops, seminars, and training sessions aimed at various demographics, from school children to adults, thereby promoting lifelong learning. These organizations also engage in advocacy, influencing policy changes that support environmental education and sustainable practices at local, national, and international levels. Moreover, NGOs frequently develop and disseminate educational materials such as books, videos, and online content, which can be used by educators and community leaders to enhance their teaching efforts.

NGOs also play a vital role in community-based projects that offer practical experiences in environmental conservation. These projects not only improve local environments but also provide participants with hands-on learning opportunities that reinforce theoretical knowledge. By involving community members in activities such as tree planting, habitat restoration, and wildlife monitoring, NGOs help foster a sense of ownership and responsibility towards the environment.

In summary, the combined efforts of educators, community leaders, and NGOs create a robust framework for promoting lifelong environmental learning. Educators lay the foundational knowledge and inspire sustainable behaviours; community leaders mobilize and engage local populations; and NGOs provide the resources, expertise, and advocacy needed to support and expand these efforts. Through their collaborative work, these groups help cultivate an informed and active citizenry dedicated to preserving the environment for future generations.

2.4 References

1. *Bamford, H., & Loughlin, C. (2020). Environmental Education: The Essential Guide. Routledge.*
2. *Bell, A. (2019). Lifelong Learning for Sustainable Development. Springer.*
3. *Carson, R. (1962). Silent Spring. Houghton Mifflin Harcourt.*
4. *Eames, C., & Williams, R. (2021). Integrating Environmental Education into the Curriculum. Cambridge University Press.*
5. *Gardner, P., & Stern, P. (2020). Environmental Literacy in Science and Society: From Knowledge to Decisions. MIT Press.*
6. *Gough, A. (2019). Education for Sustainability: The Role of Lifelong Learning. Palgrave Macmillan.*
7. *Kline, J. (2022). Community-Based Environmental Education Programs. Academic Press.*
8. *Leach, J., & Rawsthorne, M. (2023). Environmental Stewardship: Theory and Practice. Wiley-Blackwell.*
9. *Orr, D. (1992). Ecological Literacy: Education and the Transition to a Postmodern World. State University of New York Press.*
10. *Robottom, I., & Hart, P. (2021). Research in Environmental Education: Engaging the Community. Routledge.*
11. *Tilbury, D. (2022). Education and Sustainability: The Role of Lifelong Learning. Routledge.*
12. *UNESCO. (2021). Education for Sustainable Development Goals: Learning Objectives. United Nations Educational, Scientific and Cultural Organization.*
13. *Wright, T. S. A. (2022). The Role of NGOs in Environmental Education and Advocacy. Oxford University Press.*

Chapter 3: Sustainability: Concepts & Practices

Sustainability in environmental contexts is a comprehensive concept aimed at reconciling human development with the planet's ecological limits. At its heart, sustainability is about fulfilling present needs without jeopardizing the ability of future generations to meet their own needs. This idea, articulated by the Brundtland Commission in 1987, underscores the balance between economic growth, social equity, and environmental protection.

The definition of sustainability extends beyond mere resource conservation; it involves creating systems that are ecologically viable, economically viable, and socially equitable. This integrated approach recognizes the interdependencies between human activities and natural ecosystems, advocating for solutions that support environmental health, economic prosperity, and social well-being.

Key principles of sustainability include intergenerational equity, which emphasizes responsible resource management to ensure long-term availability, and the precautionary principle, which advocates for proactive measures to prevent environmental harm even in the absence of full scientific certainty. Maintaining biodiversity and ecosystem integrity is crucial, as these elements are essential for ecosystem resilience and service provision. Additionally, inclusivity ensures that diverse stakeholders, including marginalized communities, are involved in decision-making processes, promoting social justice and equitable resource distribution.

By embracing these principles, sustainability aims to foster a harmonious relationship between human societies and the natural environment, challenging traditional development models and promoting practices that safeguard the planet for future generations.

3.1 Definition & Principles of Sustainability in Environmental Contexts:

Sustainability in environmental contexts is a multifaceted concept that seeks to harmonize the demands of human development with the ecological limits of the planet. At its core, sustainability is about meeting the needs of the present without compromising the ability of future generations to meet their own needs. This principle, famously articulated by the *Brundtland Commission in 1987*, emphasizes the importance of balancing economic, social, and environmental considerations.

The definition of sustainability extends beyond simple conservation of resources. It involves creating systems that are not only ecologically viable but also economically and socially equitable. Sustainable development requires an integrated approach that takes into account the complex interdependencies between human activities and natural ecosystems. This holistic view recognizes that environmental health, economic prosperity, and social well-being are interconnected and mutually reinforcing.

One of the key principles of sustainability is the idea of intergenerational equity. This principle asserts that we have a responsibility to manage resources in a way that does not deplete them for future generations. It calls for prudent use of natural resources, ensuring that their benefits are available to people over the long term. This involves not only conserving resources but also restoring and regenerating natural systems that have been degraded.

Another crucial principle is the precautionary approach, which advocates for proactive measures to prevent environmental harm even when scientific certainty is not fully established. This principle is particularly relevant in contexts where environmental damage could be irreversible or have widespread consequences. It underscores the importance of caution and foresight in environmental management and policy-making.

Sustainability also emphasizes the importance of maintaining biodiversity and ecosystem integrity. Biodiversity, the variety of life on Earth, is critical for the resilience of ecosystems and their ability to provide essential services such as clean air, water, and food. Protecting and restoring natural habitats, reducing pollution, and mitigating climate change are all essential actions to preserve biodiversity.

The principle of inclusivity is also central to sustainability. This involves engaging a wide range of stakeholders, including marginalized and indigenous communities, in decision-making processes. Inclusivity ensures that diverse perspectives are considered and that the benefits of sustainable development are shared equitably. It recognizes the importance of social justice and the need to address issues such as poverty, inequality, and access to resources.

In practice, sustainability requires the adoption of sustainable practices across various sectors. For example, in agriculture, this might involve techniques that enhance soil health, reduce chemical inputs, and conserve water. In energy, it could mean shifting from fossil fuels to renewable sources like solar and wind power. In industry, sustainability might focus on reducing waste, increasing energy efficiency, and promoting the circular economy, where products are designed for reuse and recycling, minimizing waste and resource consumption.

Ultimately, the goal of sustainability is to create a harmonious relationship between human societies and the natural environment. It challenges us to rethink traditional models of development and to innovate in ways that respect ecological boundaries and promote social well-being. By embracing the principles of sustainability, we can work towards a future where economic growth, environmental protection, and social equity go hand in hand, ensuring a healthy and vibrant planet for generations to come.

3.2 Sustainable Development Goals & Their Relevance to Global Environmental Challenges

In September 2000, the United Nations gathered political leaders from 189 countries to announce the Millennium Development Goals (MDGs). Among these eight goals, Goal 7 was dedicated to ensuring environmental sustainability, highlighting the importance of integrating environmental considerations into global development agendas.

The MDGs period concluded in 2015, marking a transition to a new set of global objectives known as the Sustainable Development Goals (SDGs). The SDGs consist of 17 goals that address a broad range of global challenges and are intended to be achieved by the end of 2030. Of these 17 goals, nine specifically focus on various aspects of environmental protection and sustainability. These goals are designed to tackle critical issues such as:

- **Water and Sanitation (Goal 6):** We all know that many regions in different countries of the world facing water scarcity and sanitation problems. Whereas, according to World Health Organisation (WHO), one out of the three humans does not have access to safe drinking water facility. Therefore, SGD – 6 ensures availability and sustainable management of water and sanitation for all. This goal also emphasizes the need to improve water quality, manage water resources efficiently, and ensure universal access to clean drinking water and sanitation facilities.
- **Sustainable Economic Growth (Goal 8):** As per the International Labour Organisation's (ILO) report published in January 2024, global unemployment ratio stands at around 5.2%. This is a huge number and a big hurdle in the way of SDG – 8 that talks about sustainable

economic growth. On the other hand, Asia and Africa are facing some serious economic growth problems due to high poverty index and unemployment ratio. Keeping in view the above mentioned facts, SDG – 8 focuses to promotes sustained, inclusive, and sustainable economic growth, full and productive employment, and decent work for all. This goal integrates environmental considerations into economic development, advocating for growth that does not compromise environmental health.

- **Inclusive and Sustainable Industrialization (Goal 9):** Many post-colonial countries of Asia and Africa are lacking industrialization to meet the basic necessities. As a result of this lack, these countries are now being considered as non-inclusive and marginalised communities in the world. To address this issue, SDG – 9 builds resilient infrastructure, promote inclusive and sustainable industrialization, and foster innovation. This goal encourages industries to adopt sustainable practices, reduce their environmental impact, and contribute to economic growth.
- **Sustainable Human Settlements (Goal 11):** There are many thickly populated cities in the world. Cities like Beijing, Mumbai, Karachi,

Dhaka, Shanghai, Chongqing, Istanbul, Tokyo, Cairo, New York, London, and Moscow have more than 10 million populations living inside the boundaries of these thickly populated cities. These cities not only emit huge amount of energy but also create huge waste on daily basis. On the other hand, tall commercial and residential buildings pose a great threat to nature and human environment. SDG – 11 stresses to make these cities including many other cities all around the world and human settlements inclusive, safe, resilient, and sustainable. This goal also stresses to involving and improving urban planning as well as management to address issues such as pollution, waste management, and climate resilience.

- **Combating Climate Change (Goal 13):** Temperature is raising high at an alarming rate and many hot regions of the world are already experiencing severe climate change effects. Mountains covered with snow are melting including Antarctica's snow icebergs. This too much melt of snow is raising water level inside oceans including humidity level due to evaporation of water vapour. To cope up this issue, SDG – 13, calls upon the world that there is a dire need to take urgent actions to combat climate change and its impacts. This goal also aims to strengthen resilience and adaptive capacities to climate-related hazards and natural disasters.

- **Life below Water (Goal 14):** SDG – 14 highlights the need to conserve and sustainably use the oceans, seas, and marine resources. This goal focuses on preventing and reducing marine pollution, protecting marine ecosystems, and ensuring sustainable fishing practices.
- **Life on Land (Goal 15):** SDG – 15 talks about protection, restoration, and promotion of sustainable use of terrestrial ecosystems, sustainably manage forests, combat desertification, and halt and reverse land degradation and halt biodiversity loss. This goal underscores the need to safeguard terrestrial ecosystems and biodiversity.
- **Responsible Consumption and Production (Goal 12):** According to World Food Programme (WFP) organisation's report, 1.3 billion tons of food is being wasted every year. This means that 30% or around one third of the total produced food is wasted which is very much shameful for all of us. According to UN's report on global hunger, around 828 million people across the globe are being affected by

hunger due to lack of proper food. Therefore, if we can reduce the above mentioned 1.3 billion tons of food waste, we can surely provide some food to eat to these 828 million people in different corners of the world. Ensure sustainable consumption and production patterns. To cope up this issue, SDG – 12 addresses the need for efficient use of resources, reduction of waste, and promotion of sustainable practices across all sectors of society.

- **Partnerships for the Goals (Goal 17):** SDG – 17 strengthens the means of implementation and revitalize the Global Partnership for Sustainable Development. While not exclusively environmental, this goal underscores the importance of partnerships and cooperation in achieving all the SDGs, including those related to environmental sustainability.

Despite these ambitious goals, the United Nations faces significant challenges in advancing environmental protection. These challenges stem from various sources:

- ***Funding Constraints:*** We all know that adequate financial resources are so crucial in implementing and monitoring environmental initiatives. Many countries struggle with limited budgets and economic constraints, hindering their ability to meet SDG targets.

- ***Political Will and Governance Issues:*** Effective environmental protection requires strong political commitment and governance frameworks. In some regions, political instability, corruption, and lack of enforcement mechanisms undermine progress.
- ***Technological and Capacity Limitations:*** Developing countries often lack the technology and capacity necessary to implement sustainable practices and manage environmental resources effectively.
- ***Global Inequality:*** Disparities between developed and developing nations can create imbalances in the ability to address environmental challenges. Wealthier countries may have more resources to invest in sustainable solutions, while poorer countries may struggle to address basic needs.
- ***Climate Change and Environmental Degradation:*** The increasing impacts of climate change, such as extreme weather events and rising sea levels, pose severe risks to achieving environmental SDGs. Additionally, on-going environmental degradation, such as

deforestation and pollution, complicates efforts to protect ecosystems and biodiversity.

These goals and challenges underscore the critical need for coordinated global efforts, innovation, and substantial investment to address environmental issues and achieve a sustainable future for all.

3.3 Case Studies: Successful Sustainability Initiatives across Various Sectors

Across the globe, several countries have made significant strides in reducing carbon emissions, providing valuable case studies for other nations striving to achieve similar sustainability goals. These success stories not only showcase effective strategies and innovative practices but also offer a beacon of hope and a roadmap for global efforts to combat climate change. Here, we explore some notable examples of countries that have made substantial progress in their sustainability initiatives:

- **Denmark: Leading the Way in Renewable Energy:** Denmark has emerged as a global leader in renewable energy, particularly wind power. By investing heavily in wind turbine technology and supporting policies that encourage renewable energy deployment,

Denmark has succeeded in transforming its energy sector. As of 2023, wind energy accounts for over 50% of Denmark's electricity consumption, positioning the country as a model for integrating renewable energy into a national grid (Danish Energy Agency, 2023). The Danish government has set ambitious goals to become carbon-neutral by 2050, with on-going initiatives to expand offshore wind farms and develop energy-efficient technologies (Danish Ministry of Climate, Energy, and Utilities, 2024).

- **Germany: Pioneering the Energiewende:** Germany's "Energiewende" (energy transition) is a comprehensive strategy aimed at reducing greenhouse gas emissions and transitioning to renewable energy sources. Since the early 2000s, Germany has implemented policies to phase out nuclear energy, increase the share of renewables, and improve energy efficiency. The country has seen a substantial decrease in carbon emissions, with renewable energy sources accounting for around 45% of its electricity generation by 2022 (German Federal Ministry for Economic Affairs and Energy, 2023). Germany's focus on energy efficiency in buildings, expansion of solar and wind power, and commitment to climate protection policies serve as significant examples of how national policy can drive environmental progress.

- **Sweden: Achieving Carbon Neutrality:** Sweden has set an ambitious target to become one of the world's first carbon-neutral countries by 2045. The Swedish government has implemented a broad range of policies to reduce emissions, including a carbon tax, investments in green technology, and incentives for renewable energy (Swedish Environmental Protection Agency, 2023). Sweden's commitment to sustainable transportation, including a significant increase in electric vehicle usage and the expansion of public transit, has further contributed to its carbon reduction efforts. The country has also prioritized sustainable forestry and efficient use of natural resources, demonstrating a holistic approach to environmental stewardship.
- **Costa Rica: Harnessing Renewable Energy:** Costa Rica stands out for its successful implementation of renewable energy solutions. Over 99% of the country's electricity comes from renewable sources, primarily hydroelectric power, with increasing contributions from wind, solar, and biomass (Costa Rican Electricity Institute, 2023).

Costa Rica's commitment to environmental sustainability is reflected in its efforts to preserve biodiversity, protect rainforests, and promote eco-tourism. The country has also set a goal to decarbonize its economy by 2050, further illustrating its dedication to sustainability.

- **New Zealand: Focus on Climate Resilience:** New Zealand has made notable progress in climate resilience and sustainability through a range of initiatives. The country aims to achieve net-zero carbon emissions by 2050 and has introduced legislation, such as the Zero Carbon Act, to guide its climate policies (New Zealand Ministry for the Environment, 2023). New Zealand's approach includes transitioning to renewable energy, enhancing biodiversity conservation, and preparing communities for climate impacts. The country's focus on climate resilience, combined with its active engagement in international climate agreements, highlights its leadership in addressing environmental challenges.

These case studies illustrate how diverse strategies can effectively contribute to reducing carbon emissions and advancing sustainability goals. By learning from these examples, other nations can develop tailored approaches to address their unique environmental challenges and move towards a more sustainable future.

3.4 References

1. *Costa Rican Electricity Institute. (2023). Renewable Energy in Costa Rica. Retrieved from [ICE](https://www.grupoice.com/)*
2. *Costanza, R., Daly, H. E., & Bartholomew, J. A. (1991). Goals, Agenda, and Policy Recommendations for Ecological Economics. In R. Costanza (Ed.), Ecological Economics: The Science and Management of Sustainability. Columbia University Press.*
3. *Danish Energy Agency. (2023). Wind Power in Denmark. Retrieved from [Danish Energy Agency](https://ens.dk/en/our-services/statistics-data-facts/wind-power-denmark)*
4. *Danish Ministry of Climate, Energy, and Utilities. (2024). Denmark's Climate Action Plan. Retrieved from [Ministry of Climate, Energy, and Utilities](https://en.kefm.dk/)*
5. *German Federal Ministry for Economic Affairs and Energy. (2023). Energy Transition in Germany. Retrieved from [BMWi](https://www.bmwi.de/Redaktion/EN/Dossier/energy-transition.html)*

6. *New Zealand Ministry for the Environment. (2023). New Zealand's Climate Change Response. Retrieved from [MfE](https://www.mfe.govt.nz/)*
7. *Swedish Environmental Protection Agency. (2023). Sweden's Climate Policy. Retrieved from [Swedish EPA](https://www.naturvardsverket.se/Climate-Change)*
8. *United Nations Environment Programme. (2023). Environmental Challenges and the SDGs. Retrieved from [UNEP](https://www.unep.org/)*
9. *United Nations. (2000). Millennium Development Goals. Retrieved from [UN Millennium Development Goals](https://www.un.org/millenniumgoals/)*
10. *United Nations. (2015). Sustainable Development Goals. Retrieved from [UN Sustainable Development Goals](https://sdgs.un.org/goals)*
11. *United Nations. (2015). Transforming Our World: The 2030 Agenda for Sustainable Development. United Nations.*
12. *World Commission on Environment and Development (WCED). (1987). Our Common Future. Oxford University Press.*

Chapter 4: Foundations of Environmental Education

Environmental awareness and education evolved significantly during the latter half of the 20th century, driven by growing recognition of ecological issues. The modern environmental movement began in the 1960s and 1970s with pivotal events such as the publication of Rachel Carson's "Silent Spring" (1962), which highlighted the dangers of pesticides and ignited public concern about environmental health. This era also saw the first Earth Day in 1970 and the 1972 Stockholm Conference, which were crucial in establishing environmental awareness and policy frameworks globally.

However, many developing countries remained focused on industrial growth and technological advancement, often neglecting environmental concerns. Rapid industrialization in nations like India and Brazil during this period led to increased pollution and deforestation, reflecting a prioritization of economic development over ecological sustainability.

The latter part of the 20th century marked a shift towards integrating environmental education into policy and curricula. The Brundtland Commission's 1987 report, "Our Common Future," emphasized sustainable development and called for a stronger focus on environmental education.

This period laid the groundwork for future environmental initiatives, setting the stage for a more coordinated global approach to sustainability and ecological stewardship.

4.1 Historical Perspectives on Environmental Awareness & Education

Most of the time, human society has adopted a policy to be reactive rather than proactive approach to tackle different problems. For instance, we realised after the loss of almost half of the human population during World War I & II that there should be a centralised organisation to manage political and socio-economic tensions. Eventually, we formed the United Nations Organisation in 1945. As far as environmental problem is concerned, we have also adopted the same approach. Actually, environmental awareness and education began gaining traction in the latter part of the 20th century, as the global community started to recognize the pressing need to address environmental degradation and promote sustainable practices. Despite these early efforts, meaningful progress was slow, and many developing nations were primarily focused on industrial growth and technological advancements.

- **Early Beginnings:** The modern era of environmental awareness can be traced back to the 1960s and 1970s when scientific findings began to draw attention to the negative impacts of industrialization and human activities on the environment. The publication of Rachel Carson's seminal work, "Silent Spring" (1962), played a pivotal role in raising public consciousness about the dangers of pesticides and environmental pollution. Carson's book highlighted the interconnectedness of ecosystems and the potential consequences of human actions on nature, sparking widespread debate and laying the groundwork for the environmental movement.
- **The Rise of Environmental Movements:** The 1970s marked a turning point with the establishment of key environmental milestones. The first "*Earth Day*", celebrated on April 22, 1970, mobilized millions of people around the world to advocate for environmental protection and sustainability. This event was instrumental in catalysing policy changes and fostering a global environmental consciousness. The United Nations Conference on the Human Environment, held in Stockholm in 1972, further emphasized the importance of integrating environmental considerations into global

policy-making and led to the creation of the United Nations Environment Programme (UNEP, 2024).

- **Challenges in Developing Countries:** Despite these advances, many developing countries in the latter half of the 20th century were heavily focused on economic development and industrialization. The pursuit of technological innovations and industrial advantage often took precedence over environmental concerns. Nations in Asia, Africa, and Latin America were striving to improve their economic standing and compete on a global scale, leading to rapid industrial growth that frequently came at the expense of environmental health. The emphasis on economic development often overshadowed the implementation of environmental awareness and educational programs, resulting in limited progress in these regions.
- For instance, countries like India and Brazil experienced significant industrial expansion during this period. While this growth contributed to economic progress, it also led to increased pollution, deforestation, and environmental degradation. These nations were grappling with the dual challenge of fostering economic development while addressing the environmental consequences of industrialization.

- **The Shift towards Environmental Education:** As the 20th century drew to a close, there was a growing recognition of the need for environmental education as a critical component of sustainable development. The Brundtland Commission's report, "Our Common Future" (1987), introduced the concept of "sustainable development" and emphasized the importance of environmental education in achieving long-term ecological balance (World Commission on Environment and Development, 1987). This report laid the foundation for integrating environmental education into curricula and policy frameworks, advocating for a holistic approach to addressing environmental issues.

As a whole, while the latter half of the 20th century marked the beginning of a global movement towards environmental awareness and education, the pace of progress was uneven. Many developing countries were focused on industrial growth and technological advancement, often at the expense of environmental considerations. However, as the global community increasingly recognized the importance of sustainability, efforts to integrate environmental education into policy and practice began to gain momentum, setting the stage for more concerted action in the 21st century. We need to understand that natural environment is the source of our very existence and hence without protecting our environment, existence of life will not be possible in future.

4.2 Evolution of Environmental Movements & Their Impact on Policy & Public Awareness

As the 21st century began, there was a notable shift in global consciousness regarding environmental issues. This period marked a significant evolution in environmental movements, which increasingly influenced policy and public awareness. The early years of the century saw a growing realization of the importance of environmental education and a surge in campaigns aimed at preserving the planet.

- **Rise of Environmental Awareness:** In the first decade of the 21st century, the urgency of addressing environmental challenges became more evident. The impacts of climate change, such as rising global temperatures, extreme weather events, and increasing environmental degradation, underscored the need for enhanced environmental education and awareness. Key events and initiatives during this period played a crucial role in shaping public perception and policy responses.

One of the landmark events was the publication of the Intergovernmental Panel on Climate Change (IPCC) Fourth Assessment Report in 2007. This comprehensive report provided a detailed analysis of climate change impacts and solidified the scientific consensus on the need for urgent action. The IPCC's findings helped to galvanize both public and political support for environmental initiatives (IPCC, 2007). However, we still need to go far away to create a sense of awareness among 8 billion people living in different parts of the world. It should be everybody's concern as we all are being affected from the severe climate changes.

- **Increased Focus on Environmental Education:** The early 2000s saw a concerted effort to integrate environmental education into school curricula and public outreach programs. Recognizing the need to cultivate an environmentally conscious generation, many educational institutions began to incorporate sustainability topics into their teaching. Programs aimed at raising awareness about environmental issues, such as the importance of recycling, conservation, and sustainable living, gained prominence.

Initiatives like the United Nations Decade of Education for Sustainable Development (2005-2014) were instrumental in promoting environmental education worldwide. This decade-long initiative sought to integrate principles of sustainability into all levels of education and encourage actions that would lead to a more sustainable future (UNESCO, 2005).

- **Expansion of Environmental Campaigns:** The 2000s also saw a proliferation of environmental campaigns that aimed to address specific issues and mobilize public action. High-profile campaigns such as the "*350.org*" movement, founded in 2008, focused on reducing atmospheric carbon dioxide levels to a safe threshold. This campaign utilized grassroots organizing, social media, and global events to raise awareness and advocate for climate action (McKibben, 2008).

Another notable initiative was the "Earth Hour," first organized by the World Wide Fund for Nature (WWF) in 2007. This global event encouraged individuals and organizations to turn off non-essential lights for one hour to raise awareness about climate change and the need for energy conservation. Earth Hour quickly became a symbol of global solidarity in the fight against environmental issues (WWF, 2007).

- **Policy Influence & Legislative Advances:** The increased public awareness and activism of the 2000s had a significant impact on environmental policy. Governments and international organizations began to incorporate environmental concerns into their policy frameworks more robustly. The adoption of the Paris Agreement in 2015 at the 21st Conference of the Parties (*COP21*) to the UN Framework Convention on Climate Change (UNFCCC) represented a landmark moment in international climate policy. The agreement set ambitious targets for limiting global warming and encouraged countries to submit nationally determined contributions (NDCs) to reduce emissions (UNFCCC, 2015).

Furthermore, many countries began to implement stricter environmental regulations and promote sustainable practices within their jurisdictions. Policies supporting renewable energy, emissions reductions, and conservation efforts became more widespread, reflecting the growing recognition of the need for comprehensive environmental strategies.

In a nutshell, the first decade of the 21st century marked a pivotal era in the evolution of environmental movements. The increased focus on environmental education, the expansion of awareness campaigns, and the influence on policy and legislation highlighted a growing global commitment to addressing environmental challenges. These developments set the stage for continued efforts toward sustainability and reinforced the critical role of education and activism in shaping a more sustainable future.

4.3 Importance of Integrating Environmental Education into Formal, Informal, & Non-Formal Learning Settings

In today's rapidly evolving world, there is a growing recognition of the urgent need to incorporate environmental education across all learning environments. Integrating environmental education into formal, informal, and non-formal education systems is crucial for fostering a generation that is knowledgeable about environmental issues and equipped to tackle them effectively. This comprehensive approach to environmental education ensures that individuals at all stages of life are engaged in understanding and addressing ecological challenges.

- **Formal Education:** Formal education refers to structured learning that takes place in traditional academic settings such as schools and universities. Integrating environmental education into formal curricula is essential for developing foundational knowledge and critical thinking skills related to environmental issues from an early age. This integration can involve incorporating environmental topics into science, geography, and social studies courses, as well as creating dedicated environmental science programs.

  For instance, many countries have started embedding environmental education into their national education standards. In the United States, the Next Generation Science Standards (NGSS) include environmental and sustainability topics, emphasizing the importance of understanding human impact on the environment (NGSS Lead States, 2013). Similarly, the United Nations Educational, Scientific and Cultural Organization (UNESCO) has developed guidelines to help countries integrate sustainability into education systems, aiming to prepare students for the challenges of a changing world (UNESCO, 2017).

- **Informal Education:** Informal education occurs outside of formal academic institutions and includes activities such as learning through media, community programs, and personal experiences. Informal environmental education plays a significant role in shaping public awareness and behaviours related to environmental conservation. Activities such as nature walks, educational documentaries, and public service announcements contribute to a broader understanding of environmental issues.

Community organizations and NGOs often lead informal education efforts by offering workshops, seminars, and outdoor educational programs. For example, the Nature Conservancy provides various resources and activities aimed at raising environmental awareness and encouraging sustainable practices among diverse audiences (The Nature Conservancy, 2022). Such programs can effectively complement formal education by reaching individuals in more relaxed and accessible settings.

- **Non-Formal Education:** Non-formal education encompasses organized educational activities that occur outside traditional formal education but still have structured goals and objectives. This includes adult education programs, vocational training, and community-based learning initiatives. Non-formal environmental education can address specific community needs, such as sustainable agricultural practices, waste management, and renewable energy use.

  Programs such as community-based recycling initiatives and sustainability workshops often fall under non-formal education. For instance, the Green Teacher initiative offers professional development and educational resources to help educators integrate environmental topics into their teaching, thereby reaching a wider audience including both students and adults (Green Teacher, 2023). These programs are vital for lifelong learning and for equipping individuals with practical skills for sustainable living.

Integrating environmental education into formal, informal, and non-formal learning settings is essential for fostering a comprehensive understanding of environmental issues and promoting sustainable practices. This holistic approach ensures that individuals of all ages and backgrounds are informed, engaged, and empowered to contribute to environmental conservation efforts. By enhancing environmental education across various learning environments, societies can better address the complex challenges of our time and work towards a more sustainable future.

4.4 References

1. *Carson, R. (1962). "Silent Spring". Houghton Mifflin Harcourt.*
2. *Green Teacher. (2023). "Professional Development for Educators". Retrieved from [Green Teacher](https://greenteacher.com/)*
3. *Intergovernmental Panel on Climate Change (IPCC). (2007). "Climate Change 2007: The Physical Science Basis". Cambridge University Press.*
4. *McKibben, B. (2008). "The End of Nature". Random House.*
5. *NGSS Lead States. (2013). "Next Generation Science Standards: For States, By States". The National Academies Press.*
6. *Sachs, J. D. (2005). "The End of Poverty: Economic Possibilities for Our Time". Penguin Books.*
7. *The Nature Conservancy. (2022). "Education and Outreach". Retrieved from [Nature Conservancy](https://www.nature.org/en-us/what-we-do/our-insights/perspectives/education-and-outreach/)*
8. *United Nations Educational, Scientific and Cultural Organization (UNESCO). (2005). "United Nations Decade of Education for Sustainable Development (2005-2014)". Retrieved from [UNESCO](https://en.unesco.org/themes/education-sustainable-development)*

9. *United Nations Educational, Scientific and Cultural Organization (UNESCO). (2017). "Education for Sustainable Development Goals: Learning Objectives". Retrieved from [UNESCO](https://unesdoc.unesco.org/ark:/48223/pf0000247444)*
10. *United Nations Environment Programme (UNEP). (2024). "The Origins and Evolution of UNEP". Retrieved from [UNEP](https://www.unep.org/about-un-environment)*
11. *United Nations Framework Convention on Climate Change (UNFCCC). (2015). "Paris Agreement". Retrieved from [UNFCCC](https://unfccc.int/process-and-meetings/the-paris-agreement/the-paris-agreement)*
12. *World Commission on Environment and Development. (1987). "Our Common Future". Oxford University Press*
13. *World Wide Fund for Nature (WWF). (2007). "Earth Hour: A Global Movement". Retrieved from [WWF](https://www.earthhour.org/)*

Chapter 5: Synthetic Biology & Its Environmental Implications

Synthetic Biology is a cutting-edge field that integrates biology and engineering to design and construct new biological entities or reengineer existing ones for practical purposes. This interdisciplinary approach combines genetics, molecular biology, and bioengineering to address complex problems and develop innovative solutions across various sectors. Its applications are particularly promising in environmental sustainability.

A key focus of synthetic biology is the textile industry, where traditional synthetic fibres like polyester and nylon, derived from fossil fuels, significantly contribute to environmental degradation. For instance, these fibres account for about 10% of global carbon emissions and contribute to landfill waste due to their short lifecycle. Scientists like Stephen Wallace advocate for synthetic biology's potential to revolutionize textile production by developing sustainable alternatives, such as biodegradable fibres and bio-fabricated materials.

Synthetic biology's role in environmental protection extends beyond textiles. Researchers are engineering microorganisms to address pollution and waste management challenges. Examples include genetically modified bacteria that can clean up oil spills or produce biofuels and biodegradable plastics from renewable resources. These innovations offer sustainable solutions to critical environmental issues.

Overall, synthetic biology holds transformative potential for enhancing environmental sustainability by creating new materials and processes that reduce our ecological footprint. Continued research and investment are crucial to realizing its full benefits and addressing environmental challenges effectively.

5.1 Introduction to Synthetic Biology & Its Applications

Synthetic Biology which is one the recently emerged branches of Biology, represents a ground-breaking field that merges Biology and Engineering to create new biological parts, systems, and organisms, or to redesign existing biological systems for useful purposes. This interdisciplinary domain harnesses principles from various scientific disciplines, including Genetics, Molecular Biology, and Bioengineering, to address complex problems and innovating solutions across diverse sectors. In the case of environmental protection for sustainable futures, synthetic Biology has injected a soul and reflected a ray of hope. However, this recently developed branch requires lots of research and financial investments. In this regard, many leading universities of the world particularly Western universities have already started working on it to lead the world in this newly emerged field of biological sciences.

- **Role of Synthetic Fibres in the Clothing Industry:** In a TEDx Talk delivered on April 10, 2023, a British origin scientist, named Stephen Wallace (Synthetic Biology Expert) highlighted the significant environmental impact of synthetic fibres being used in clothing. According to Wallace, modern apparel is often composed of both natural and synthetic fibres. The synthetic fibres, which are a major component of many garments, are manufactured in factories using fossil fuels. Specifically, about two-thirds of the production process involves fossil fuels, contributing to environmental degradation and resource depletion (Wallace, 2023).

  These synthetic fibres, such as polyester and nylon, are derived from petrochemical sources and hence require a considerable amount of energy to produce. Consequently, the clothing industry is a major contributor to global carbon dioxide emissions, responsible for approximately 10% of total emissions (Global Fashion Agenda, 2023). Furthermore, the lifecycle of these garments is relatively short; they typically end up in landfills within five to ten years of use. This short lifespan, combined with the environmental footprint of their production, underscores the need for more sustainable practices in the fashion industry.

- **Impact of Synthetic Biology on Sustainable Solutions:** As discussed earlier, synthetic Biology offers promising approaches to mitigate the environmental impact of traditional textile production. By employing synthetic Biology techniques, researchers are developing innovative methods to produce sustainable materials and reduce dependency on fossil fuels. For example, bioengineered materials such as bio-fabricated fabrics made from microorganisms or renewable resources present a viable alternative to conventional synthetic fibres.

  One notable application is the development of "biodegradable" synthetic fibres produced from bio-based sources. These fibres aim to mimic the properties of traditional synthetic materials while being designed to decompose more readily in natural environments, thus reducing landfill waste. Additionally, synthetic Biology is being used to engineer microorganisms that can produce natural fibres like silk or spider silk, which have potential applications in textiles and other industries (Kumar et al., 2022).

Overall, the integration of synthetic Biology into textile manufacturing presents an opportunity to address some of the most pressing environmental challenges posed by traditional clothing production. By developing new, more sustainable materials and reducing reliance on fossil fuels, synthetic Biology has the potential to significantly lower the clothing industry's carbon footprint and reduce the environmental impact of textiles. As the industry continues to evolve, embracing these innovative solutions will be crucial for achieving greater sustainability in fashion and beyond.

5.2 Environmental Risks & Ethical Considerations Associated with Synthetic Biology

Synthetic Biology, a rapidly advancing field at the intersection of Biology and engineering, holds immense promise for solving some of the world's most pressing challenges, from sustainable production to medical advancements. However, this technology also introduces a range of environmental risks and ethical considerations that must be carefully managed to ensure responsible development and application. Let us now discuss these environmental risks and ethical considerations one by one.

- **Biosafety Concerns:** One of the primary environmental risks associated with synthetic Biology is the potential for unintended consequences resulting from the release of genetically engineered organisms into the environment. These organisms, designed to perform specific functions, may interact unpredictably with natural ecosystems. For example, engineered microbes intended to degrade pollutants might disrupt local microbial communities or unintentionally produce harmful by-products (Gordon et al., 2019).
- **Gene Flow & Biodiversity:** The risk of gene flow from synthetic organisms to wild relatives is another significant concern. Engineered genes may escape into natural populations, potentially leading to ecological imbalances. For instance, synthetic organisms designed to enhance agricultural productivity could crossbreed with wild species, potentially altering ecosystems and affecting biodiversity (Kumar et al., 2021).

- **Waste & Resource Use:** The production and disposal of synthetic biological materials also pose environmental challenges. The manufacturing processes for synthetic Biology products often involve substantial energy and resource use, which can contribute to environmental degradation. Additionally, the disposal of synthetic biological materials may present risks if these materials do not degrade as expected or if they persist in the environment in ways that are difficult to predict (Anderson et al., 2014).
- **Biological Ethics & Design:** The ethical implications of designing and constructing synthetic organisms raise significant questions. There are concerns about the extent to which humans should interfere with biological systems. The creation of new life forms, even for beneficial purposes, challenges traditional views of nature and raises questions about the moral status of synthetic organisms (Buchanan et al., 2008).
- **Dual Use & Misuse:** Synthetic Biology also carries the risk of misuse. The technology has potential applications in both beneficial and harmful contexts. For instance, while synthetic Biology can be used to develop new medical treatments, it could also be exploited for

bioterrorism or the creation of harmful biological agents like COVID – 19 which put entire world on pause for about 2 years. Ensuring that synthetic Biology research is conducted responsibly and ethically is critical to mitigating these risks (Fukuyama, 2015).

- **Equity & Access:** The development and application of synthetic Biology can exacerbate existing inequalities. Access to advanced synthetic Biology technologies may be limited to wealthy nations or corporations, potentially leading to disparities in benefits and risks. Additionally, the environmental and health impacts of synthetic Biology applications may disproportionately affect marginalized communities, raising concerns about justice and equity (Sarewitz, 2018).

While synthetic Biology offers transformative potential for addressing global challenges, it is crucial to address the associated environmental risks and ethical considerations. Balancing innovation with precautionary measures and ethical considerations is essential for the responsible development of synthetic Biology. This approach will help ensure that the benefits of synthetic Biology are maximized while minimizing potential harms and addressing societal concerns.

5.3 Case Studies & Examples Illustrating the Intersection of Synthetic Biology & Environmental Sustainability

Microorganisms, that are abundant and diverse, function as nature's chemical factories. Harnessing their potential through synthetic Biology offers a promising avenue for advancing environmental sustainability. By utilizing and engineering these microorganisms, we can address environmental challenges and create sustainable solutions for future generations. However, this potential requires substantial research and resource investment to fully realize its benefits. Let us now discuss the intersection between microbial Chemistry and synthetic Biology including a couple of case studies to understand their possible use for environmental sustainability.

- **Microbial Chemistry & Its Applications:** Microbial chemistry is already an integral part of various industries. For example, microorganisms are used extensively in the production of everyday products such as cheese, yogurt, sauces, beverages, and butter. These microbes naturally ferment and process raw materials to produce these items, demonstrating their capability as biological factories (van der Meer et al., 2010).

- **Synthetic Biology as a Tool for Environmental Sustainability:** Stephen Wallace highlights that synthetic Biology represents an emerging technology with the potential to revolutionize our approach to environmental conservation. By programming microbes to perform specific tasks, we can transform them into powerful tools for environmental protection. This involves altering the metabolic pathways of microorganisms to enable them to perform functions that benefit the environment, such as bioremediation or waste management.

  For instance, researchers are developing engineered microbes to degrade pollutants or convert waste into valuable products. By modifying the DNA of these microorganisms, scientists can enhance their natural abilities or endow them with new capabilities to address environmental issues. This process, known as metabolic engineering, involves redesigning microbial metabolism to optimize their performance for specific environmental applications (Kumar et al., 2022).

- **Case Study Bioremediation:** One notable example is the use of genetically modified bacteria for bioremediation, where microbes are engineered to break down environmental contaminants such as oil spills or heavy metals. For example, researchers have developed bacteria that can degrade petroleum hydrocarbons, aiding in the clean-up of oil spills and reducing environmental damage (Rosenberg et al., 2010).
- **Case Study of Sustainable Production:** Another application is the creation of biofuels and biodegradable materials. Synthetic Biology allows for the engineering of microorganisms to produce biofuels from renewable resources, potentially reducing reliance on fossil fuels. Additionally, engineered microbes can produce bio plastics that decompose naturally, minimizing plastic waste and pollution (Gibson et al., 2016).

The intersection of synthetic Biology and environmental sustainability holds great promises. By leveraging the natural capabilities of microorganisms and enhancing them through synthetic Biology, we can develop innovative solutions for environmental challenges. While the potential is significant, realizing these benefits requires continued research and investment. By changing microbial DNA and programming these organisms to perform specific environmental tasks, we can create a more sustainable future and address pressing ecological issues effectively.

5.4 References

1. *Anderson, J. C., et al. (2014). "Synthetic Biology: Environmental Risks and Policy Considerations". Nature Reviews Genetics, 15(10), 603-612.*
2. *Buchanan, A., et al. (2008). "The Ethical Challenges of Synthetic Biology". Science and Engineering Ethics, 14(2), 245-254.*
3. *Fukuyama, F. (2015). "The Ethics of Synthetic Biology: A Policy Perspective". Policy Review, 220, 89-98.*
4. *Gibson, D. G., et al. (2016). "Synthetic Biology: Advancements and Applications in Biofuel Production". Current Opinion in Biotechnology, 37, 61-68.*
5. *Global Fashion Agenda. (2023). "Fashion and Sustainability Report". Retrieved from [Global Fashion Agenda](https://globalfashionagenda.com/)*
6. *Gordon, G., et al. (2019). "Biosafety and Environmental Risks of Synthetic Biology". BioScience, 69(6), 473-483.*
7. *Kumar, P., Singh, R., & Gupta, S. (2021). "Gene Flow and Biodiversity Risks in Synthetic Biology". Environmental Biosafety Research, 20(3), 45-58.*
8. *Kumar, P., Singh, R., & Gupta, S. (2022). "Advances in Synthetic Biology for Sustainable Materials". Journal of Sustainable Materials and Technologies, 14, 123-135.*

9. *Kumar, P., Singh, R., & Gupta, S. (2022). "Metabolic Engineering of Microorganisms for Environmental Applications". Biotechnology Advances, 55, 107-123.*
10. *Rosenberg, E., et al. (2010). "Bioremediation of Oil Spills by Microorganisms". Nature Reviews Microbiology, 8, 410-422.*
11. *Sarewitz, D. (2018). "Synthetic Biology and Social Justice: Issues of Equity and Access". Environmental Research Letters, 13(7), 074002*
12. *van der Meer, J. R., et al. (2010). "Microbial Biotransformations and Their Applications'. Annual Review of Microbiology, 64, 211-234.*
13. *Wallace, S. (2023). "TEDx Talk: The Environmental Impact of Synthetic Fibres". TEDx Vienna. Retrieved from [TEDx Talks](https://www.ted.com/tedx)*

Chapter 6: Microchemistry: Innovations & Environmental Challenges

Microchemistry, the study of chemical processes at a microscopic scale, is essential in environmental monitoring and analysis. This field focuses on detecting and analysing contaminants, tracking pollutants, and assessing environmental health with high precision and efficiency.

Microchemistry's role extends to various applications, including drug production. The pharmaceutical industry relies heavily on microchemistry for synthesizing medications, many of which are derived from fossil fuels. Approximately 70% of pharmaceuticals are produced using fossil fuel-based chemicals, contributing to significant environmental impacts due to the extensive fossil fuel consumption in their production.

The environmental impact of fossil fuels is substantial, with about 10 billion tons consumed globally each year, leading to pollution, climate change, and degradation of natural resources. This underscores the urgency of finding sustainable practices and alternatives.

Emerging solutions in microchemistry include the exploration of alternative sources for chemicals like adipic acid, commonly used in manufacturing. Innovations such as utilizing lignin, a by-product from paper milling and agricultural processes, as a raw material and synthetic biology applications for converting plastic waste into useful chemicals offer promising sustainable approaches.

Overall, microchemistry is pivotal in developing and implementing solutions for environmental challenges, promoting sustainability through innovative and less harmful practices.

6.1 Role of Microchemistry in Environmental Monitoring & Analysis

As hinted in the previous chapter, Microchemistry plays a crucial role in various fields, including environmental monitoring and analysis. It involves the study and manipulation of chemical processes at a microscopic scale, which can be pivotal in detecting and analysing environmental contaminants, tracking pollutants, and assessing environmental health. To understand it further, let us know the following important facts.

- **Microchemistry in Drug Production:** The pharmaceutical industry heavily relies on Microchemistry for drug development and production. Microchemists are instrumental in designing and synthesizing medications, including painkillers and antibiotics. However, a significant portion of these drugs is derived from fossil fuels. According to Stephen Wallace, approximately 70% of medicines, including many painkillers and antibiotics, are produced from fossil fuel-derived chemicals (Wallace, 2023).

- **Fossil Fuels & Environmental Impact:** Stephen Wallace also highlights the extensive use of fossil fuels in various industries. Annually, around 10 billion tons of fossil fuels are consumed globally for energy production and chemical manufacturing (Wallace, 2023). The reliance on fossil fuels contributes significantly to pollution and climate change. The extraction, processing, and combustion of fossil fuels release pollutants into the atmosphere, leading to greenhouse gas emissions and environmental degradation.
- The impact of fossil fuel use extends beyond climate change. It affects air and water quality, contributes to the degradation of natural habitats, and poses risks to human health. The environmental consequences underscore the need for sustainable practices and alternative technologies that minimize reliance on fossil fuels.

6.2 Environmental Impacts of Chemical Innovations & Emerging Contaminants

Chemical innovations have revolutionized many industries, including textiles and pharmaceuticals. Adipic acid $(CH_2)_4(COOH)_2$, for example, is a key ingredient used in the production of nylon and various synthetic fabrics. It is a widely utilized chemical ingredient in modern manufacturing due to its versatility and effectiveness. However, the environmental impact of adipic acid production raises significant concerns.

6.2.1 Environmental Impact of Adipic Acid Production

Adipic acid is predominantly derived from fossil fuels through a chemical process that releases substantial amounts of Carbon Dioxide. This process contributes significantly to global warming and environmental degradation, as the combustion of fossil fuels generates greenhouse gases that exacerbate climate change.

The reliance on fossil fuels for adipic acid production is problematic because it not only depletes non-renewable resources but also contributes to severe environmental issues. So an important question arises here that: Are there any alternative sources for producing adipic acid that do not rely on fossil fuels? Let us now find the answer to this important question as mentioned below.

- **Alternative Sources for Adipic Acid:** Fortunately, there are promising alternatives to fossil fuel-derived adipic acid. One such alternative is lignin $(C_{31}H_{34}O_{11})_n$ which is a complex organic polymer found in the cell walls of plants, particularly in trees. Lignin is composed of guaiacol (C_6H_4), a compound that is already produced in large quantities, approximately 17 million tons annually, through paper milling and agricultural processes. Currently, this guaiacol is often considered a waste by-product (Wallace, 2023).

  By repurposing lignin, we can explore sustainable methods for adipic acid production. For instance, combining guaiacol with adipic acid can potentially lead to more environmentally friendly chemical processes. Additionally, synthetic Biology offers innovative approaches to converting plastic waste into adipic acid, further reducing our reliance on fossil fuels and minimizing environmental impact (Wallace, 2023).

- **Synthetic Biology & Microbial Applications:** Synthetic Biology is playing a transformative role in developing sustainable chemical processes. By harnessing the power of microorganisms, researchers are exploring new ways to produce valuable chemicals and materials. For example, synthetic Biology can help convert plastic waste into useful chemicals, including adipic acid. This approach not only addresses the problem of plastic pollution but also provides a sustainable source of essential chemicals.

Moreover, Stephen Wallace points out that the human body is home to approximately 40 trillion microbes. These microorganisms have the potential to be utilized in natural medicine and various other applications. By harnessing these microbes, we can develop innovative treatments including sustainable solutions that further enhance our ability to address environmental and health challenges (Wallace, 2023). However, there is dire need to conduct more research in this particular phenomenon.

6.3 Strategies for Sustainable Chemistry Practices & Pollution Prevention

The staggering figure of 10 billion tons of fossil fuels consumed annually underscores the urgent need for sustainable Chemistry practices and pollution prevention strategies. The environmental impact of fossil fuel use—ranging from greenhouse gas emissions to ecological degradation—demands immediate and innovative action to ensure a sustainable future for coming generations. Addressing this challenge requires a multifaceted approach that includes reducing fossil fuel reliance, enhancing chemical processes, and promoting green technologies. To understand further about these multifaceted approaches, let us know some key facts as mentioned below.

6.3.1 Transition to Renewable Energy Sources

One of the most effective strategies for reducing the environmental impact of fossil fuels is to transition to renewable energy sources. This involves shifting from fossil fuels to clean energy alternatives such as solar, wind, hydroelectric, and geothermal power. These renewable energy sources produce little to no greenhouse gas emissions and hence significantly decrease the overall carbon footprint of energy production.

6.3.2 Implementing Green Chemistry Principles

Green Chemistry aims designing the chemical processes and products which minimize environmental impacts including enhancing safety measures. Green Chemistry can possibly be a new term for you and hence to make you understand in simple wording the key principles of the term Green Chemistry include:

- **Preventing Waste:** Designing processes that reduce or eliminate the generation of hazardous substances.
- **Designing Safer Chemicals:** Creating chemicals that are non-toxic and have minimal environmental impact.
- **Sustainable Synthesis:** Utilizing renewable feed stocks and energy-efficient methods to produce chemicals.

Adopting these principles in chemical manufacturing can lead to significant reductions in pollution and resource consumption.

6.3.3 Enhancing Chemical Recycling & Circular Economy

Chemical recycling involves breaking down process of the used chemicals and materials into their original components for reuse. This approach supports the circular economy model, where products and materials are continuously recycled, minimizing waste and reducing the need for new raw materials. Advancements in chemical recycling technologies can help address the challenges of plastic waste and other environmental pollutants.

- **Developing & Utilizing Alternative Materials:** The development of alternative materials that are biodegradable or derived from renewable resources can reduce dependence on fossil fuels. Examples include bio-plastics made from plant-based materials and bio-based chemicals produced through sustainable processes. These alternatives offer a way to reduce the environmental impact associated with traditional petroleum-based products.
- **Promoting Energy Efficiency:** Improving energy efficiency across various industrial processes can lead to substantial reductions in fossil fuel consumption. This includes optimizing production methods, enhancing process design, and employing energy-efficient

technologies. By increasing the efficiency of energy use, industries can lower their overall fossil fuel requirements and reduce associated emissions.

- **Supporting Research & Innovation:** Investment in research and development of new technologies and processes is crucial for advancing sustainable Chemistry practices. Innovations in areas such as green catalysts, alternative energy sources, and environmentally friendly materials can drive progress towards more sustainable chemical practices and pollution prevention.

Addressing the environmental impact of fossil fuel consumption requires a comprehensive and proactive approach. By transitioning to renewable energy sources, implementing green Chemistry principles, enhancing chemical recycling, developing alternative materials, promoting energy efficiency, and supporting research and innovation, we can make significant strides towards a more sustainable and pollution-free future. These strategies are essential for reducing our reliance on fossil fuels and protecting the environment for future generations.

6.4 References

1. *Anastas, P. T., & Warner, J. C. (1998). "Green Chemistry: Theory and Practice". Oxford University Press.*
2. *Anderson, J. R., & Davidson, D. R. (2021). "Advances in Chemical Recycling and Circular Economy". Journal of Cleaner Production, 280, 124-135.*
3. *International Renewable Energy Agency (IRENA). (2022). "Renewable Capacity Statistics 2022". Retrieved from [IRENA](https://www.irena.org/Publications/2022/June/Renewable-Capacity-Statistics-2022)*
4. *National Science Foundation (NSF). (2021). "Advancing Sustainable Chemistry through Innovation". Retrieved from [NSF](https://www.nsf.gov/funding/pgm_summ.jsp?pims_id=503469)*
5. *Ramesh, K., & Banu, J. R. (2020). "Bioplastics and Their Environmental Impact". Renewable and Sustainable Energy Reviews, 117, 109-120.*
6. *U.S. Environmental Protection Agency (EPA). (2022). "Energy Efficiency in Industry". Retrieved from [EPA](https://www.epa.gov/energy/energy-efficiency-industry)*
7. *Wallace, S. (2023). "TEDx Talk: The Environmental Impact of Fossil Fuels". TEDx Vienna. Retrieved from [TEDx Talks](https://www.ted.com/tedx)*

Chapter 7: Energy: Opportunities & Challenges

In the 21st century, global energy consumption continues to rely heavily on non-renewable sources, notably fossil fuels like coal, oil, and natural gas, as well as nuclear power. Despite advancements in technology and increased environmental awareness, these energy sources remain dominant.

Non-Renewable Energy Sources: Fossil fuels are central to electricity generation, transportation, and industry, but their combustion releases large amounts of carbon dioxide (CO_2) and other greenhouse gases, driving climate change. Coal-fired power plants are significant contributors to CO_2 emissions. Nuclear power, while generating minimal greenhouse gases, poses challenges such as radioactive waste management and safety risks associated with accidents and water use.

Environmental Consequences: The reliance on non-renewable energy has severe impacts, including climate change, which leads to rising sea levels, extreme weather events, and ecosystem disruptions. Air pollution from fossil fuels exacerbates respiratory and cardiovascular diseases and contributes to smog and acid rain. Resource depletion is a major concern as fossil fuel reserves are finite and extraction becomes more challenging. Additionally, nuclear waste presents long-term environmental hazards.

Transition to Renewable Energy: To mitigate these issues, transitioning to renewable energy sources like solar, wind, hydro, and geothermal power is essential. These alternatives offer lower environmental impacts, sustainable energy solutions, and contribute significantly to reducing greenhouse gas emissions, enhancing energy security, and protecting ecosystems.

7.1 Global Energy Consumption Trends & Their Environmental Consequences

In the 21st century, global energy consumption remains heavily dependent on non-renewable sources. This continued reliance includes fossil fuels such as coal, oil, and natural gas as well as nuclear power. Despite significant technological advancements and growing awareness of environmental issues, these non-renewable energy sources still dominate the global energy landscape. Similarly, countries like China, India, United States of America, Japan, Germany, and Brazil still heavily on fossil fuels energy. These countries are the leading carbon emitting countries in the world as well.

7.1.1 Non-Renewable Energy Sources

- **Fossil Fuels:** Fossil fuels, including coal, oil, and natural gas, are the primary sources of energy worldwide. They are used for generating electricity, powering vehicles, and supporting industrial processes. The combustion of these fuels releases considerable amounts of carbon dioxide (CO_2) and other greenhouse gases, which contribute significantly to global warming and climate change. For instance, coal-fired power plants are among the largest sources of CO_2 emissions.
- **Nuclear Power:** Nuclear power plants use nuclear fission of uranium or plutonium to produce large amounts of energy. While nuclear energy produces minimal greenhouse gas emissions compared to fossil fuels, it presents other environmental and safety challenges. These include the long-term storage of radioactive waste and the risk of catastrophic accidents, such as the Fukushima and Chernobyl disasters. Additionally, nuclear power plants require substantial water resources for cooling, which can impact local water availability.

7.1.2 Environmental Consequences

- **Climate Change:** The burning of fossil fuels releases CO_2 and other greenhouse gases into the atmosphere. These gases trap heat and lead to global warming, which in turn causes rising sea levels, more frequent and severe weather events (such as hurricanes and droughts), and disruptions to ecosystems and agriculture. The Intergovernmental Panel on Climate Change (IPCC) reports that immediate and substantial reductions in greenhouse gas emissions are required to limit global temperature rise.
- **Air Pollution:** Combustion of fossil fuels also produces various air pollutants, including sulphur dioxide (SO_2), nitrogen oxides (NOx), and particulate matter. These pollutants contribute to smog formation, acid rain, and respiratory issues in humans. Air pollution from coal power plants and vehicles is linked to increased incidences of respiratory diseases, cardiovascular problems, and premature deaths.
- **Resource Depletion:** Non-renewable energy sources are finite and will eventually be exhausted. Overreliance on these resources can lead to their depletion, which may create economic and geopolitical instability. As easily accessible reserves are depleted, extraction

becomes more challenging and costly, potentially leading to increased prices and conflicts over remaining resources.

- **Nuclear Waste:** Nuclear energy produces radioactive waste that remains hazardous for thousands of years. Safe and long-term storage of this waste is a major challenge. The disposal of radioactive materials requires highly secure facilities to prevent contamination and ensure public safety. Additionally, the potential for leaks or accidents in storage facilities poses on-going risks.
- **Environmental Degradation:** The extraction and processing of fossil fuels can lead to significant environmental damage. For example, coal mining can result in habitat destruction, water pollution, and soil erosion. Oil spills from drilling operations can have devastating effects on marine and coastal ecosystems.

7.1.3 Transition to Renewable Energy

To address the environmental and sustainability challenges associated with non-renewable energy sources, transitioning to renewable energy sources is crucial. Renewable energy options, such as solar, wind, hydro, and geothermal power, offer cleaner alternatives with lower environmental impacts. These sources are sustainable and produce little to no greenhouse gas emissions.

- **Solar Energy:** Solar power harnesses energy from the sun using photovoltaic cells or solar thermal systems. It is abundant and widely available, with the potential to significantly reduce reliance on fossil fuels.
- **Wind Energy:** Wind power uses turbines to convert wind energy into electricity. It is a growing source of renewable energy that has the potential to meet substantial portions of global energy needs.
- **Hydropower:** Hydropower generates electricity by using the energy of flowing water. While it can provide a reliable and consistent energy source, large-scale hydropower projects can have ecological impacts, such as altering river ecosystems and displacing communities.
- **Geothermal Energy:** Geothermal power utilizes heat from the Earth's interior to generate electricity. It is a stable and reliable source of energy with a low environmental footprint, though it is geographically limited to regions with significant volcanic activity.

Overall, the continued dependence on non-renewable energy sources poses serious challenges for environmental sustainability and global health. The transition to renewable energy sources is essential for mitigating the adverse effects of current energy practices and addressing climate change. This transition requires coordinated efforts from governments, industries, and individuals to promote sustainable energy solutions, enhance energy efficiency, and reduce overall environmental impacts.

7.2 Renewable Energy Technologies & Their Potential to Mitigate Climate Change

As the world particularly, South Asia and Central Asia including Europe are already grappling with the impacts of climate change, the transition to renewable energy sources has become a critical focus for mitigating environmental damage and ensuring a sustainable future. Although significant progress has been made in adopting renewable energy technologies, there is still considerable work to be done to fully realize their potential. This overview explores the various renewable energy technologies and their roles in combating climate change.

7.2.1 Key Renewable Energy Technologies

- **Photovoltaic (PV) Cells:** PV cells convert sunlight directly into electricity using semiconductor materials. Solar panels, which consist of many PV cells, are increasingly common on residential rooftops and commercial buildings. Solar energy is abundant and scalable, ranging from small-scale installations to large solar farms.
- **Solar Thermal Systems:** These systems use sunlight to heat a fluid, which can then be used to generate electricity or provide direct heating. Solar thermal power plants, also known as Concentrating Solar Power (CSP) plants, use mirrors or lenses to focus sunlight onto a small area, producing high temperatures that drive a turbine.
- **Onshore Wind Turbines:** These turbines are installed on land and use the kinetic energy of wind to generate electricity. Wind farms can be found in a variety of locations, from rural areas to offshore platforms.
- **Offshore Wind Turbines:** Installed in bodies of water, offshore wind turbines benefit from stronger and more consistent winds compared to their onshore counterparts. They have the potential to generate

significant amounts of electricity but face higher installation and maintenance costs.

- **Run-of-River Systems:** These systems divert a portion of a river's flow through turbines to generate electricity without requiring large reservoirs. They have a lower environmental impact compared to traditional hydropower projects.
- **Pumped Storage:** This method stores energy by using excess electricity to pump water uphill to a reservoir. During periods of high demand, the stored water is released to generate electricity. It acts as a form of grid energy storage, balancing supply and demand.
- **Geothermal Power Plants:** These plants use heat from the Earth's interior to generate electricity. Geothermal reservoirs are accessed by drilling wells into underground heat sources. The steam or hot water from these reservoirs drives turbines connected to electricity generators.
- **Geothermal Heat Pumps:** These systems use the stable temperatures underground to provide heating and cooling for buildings. They are highly efficient and can reduce reliance on fossil fuels for climate control.

- **Biogas:** Produced from the anaerobic digestion of organic matter, biogas can be used to generate electricity or as a fuel for vehicles. It is a renewable alternative to natural gas and can help reduce methane emissions from landfills.
- **Biofuels:** Derived from plants and organic waste, biofuels such as ethanol and biodiesel can replace gasoline and diesel. They offer a renewable alternative to fossil fuels and can help lower greenhouse gas emissions from transportation.
- **Tidal Energy:** This technology harnesses the energy from tidal movements in coastal areas. Tidal turbines and barrages capture the kinetic and potential energy from rising and falling tides to generate electricity.
- **Wave Energy:** Wave energy converters capture the energy from surface waves and use it to generate electricity. While still in the developmental stages, this technology has the potential to contribute significantly to renewable energy production. However, we need more research and successful pilot projects so that world can take benefit from this form of energy economically and effectively.

7.2.2 Potential to Mitigate Climate Change

- **Reduction of Greenhouse Gas Emissions:** Renewable energy technologies produce little to no greenhouse gases compared to fossil fuels. By replacing coal, oil, and natural gas with renewable sources, we can significantly reduce CO_2 emissions and help limit global temperature rise.
- **Energy Security and Independence:** Renewable energy sources are abundant and widely available, reducing dependence on imported fuels and enhancing energy security. They also provide opportunities for local energy production, which can stimulate economic development and job creation.
- **Sustainability and Environmental Protection:** Renewable energy technologies have lower environmental impacts than traditional energy sources. For example, wind and solar power require minimal water usage, unlike coal and nuclear power plants. Additionally, many renewable technologies have smaller footprints and less disruptive effects on ecosystems.
- **Energy Accessibility:** Renewable energy can be deployed in remote and underserved areas, providing access to electricity for communities

that lack reliable power sources. This can improve living standards and support economic development in developing regions.

- **Technological Innovation:** On-going research and development in renewable energy technologies are leading to more efficient and cost-effective solutions. Advances in energy storage, smart grids, and energy management systems are enhancing the reliability and integration of renewable energy into existing infrastructure.

7.2.3 Challenges & Future Directions

Despite the promising potential of renewable energy technologies, several challenges remain. These include:

- **Intermittency:** Many renewable sources, such as solar and wind, are intermittent and depend on weather conditions. Developing effective energy storage solutions and grid management strategies is crucial to ensure a stable energy supply.
- **Infrastructure & Integration:** Integrating renewable energy into existing power grids requires significant upgrades to infrastructure and grid management systems. Investments in smart grids and energy

storage technologies are necessary to accommodate the variability of renewable sources.

- **Economic & Policy Support:** Continued support from governments and financial institutions is needed to drive the adoption of renewable energy technologies. Policies that promote research, provide incentives, and establish clear regulatory frameworks are essential for fostering growth in the renewable energy sector.

In a nutshell, renewable energy technologies hold great promise for mitigating climate change and achieving a sustainable energy future. While significant progress has been made, further efforts are required to address challenges and accelerate the transition to a renewable energy-based economy. By investing in and advancing these technologies, we can reduce greenhouse gas emissions, enhance energy security, and protect the environment for future generations.

7.3 Policies & Initiatives Promoting Energy Efficiency & Conservation

Promoting energy efficiency and conservation is essential for reducing energy consumption, minimizing environmental impacts, and ensuring sustainable resource use. Effective policies and initiatives play a crucial role in driving these efforts. This overview explores various strategies and measures which different governments and organizations can implement to enhance energy efficiency and conservation.

7.3.1 Importance of Energy Efficiency & Conservation

- **Reduction of Energy Consumption:** Energy efficiency involves using less energy to perform the same task or provide the same service. By adopting more efficient technologies and practices, we can significantly reduce overall energy consumption.
- **Environmental Benefits:** Improving energy efficiency helps decrease greenhouse gas emissions and other pollutants, contributing to cleaner air and a lower carbon footprint.
- **Economic Savings:** Energy-efficient technologies often result in lower energy bills for businesses and households. Additionally, energy conservation can reduce the need for costly energy infrastructure investments.

7.3.2 Energy Efficiency Standards & Regulations

- **Building Codes:** Governments can implement and enforce building codes that mandate energy-efficient design and construction practices.

These codes may include requirements for insulation, windows, lighting, and HVAC systems to improve overall building performance.

- **Appliance Efficiency Standards:** Setting minimum efficiency standards for appliances and equipment ensures that new products meet high-performance criteria. For example, standards for refrigerators, washing machines, and lighting can drive the adoption of more efficient technologies.
- **Industrial Efficiency Regulations:** Policies targeting industrial processes can promote the adoption of energy-efficient technologies and practices. This may include incentives for upgrading equipment, optimizing processes, and implementing energy management systems.

7.3.3 Energy Conservation Programs & Incentives

- **Rebates and Tax Credits:** Governments and utilities can offer financial incentives such as rebates, tax credits, and grants to encourage individuals and businesses to invest in energy-efficient technologies. These incentives help offset the upfront costs of purchasing and installing energy-saving products.

- **Energy Audits:** Programs that provide free or subsidized energy audits help identify opportunities for energy savings in homes and businesses. Auditors assess energy use, recommend improvements, and estimate potential savings.
- **Demand Response Programs:** Utilities can implement demand response programs that incentivize consumers to reduce or shift their energy use during peak periods. This helps balance supply and demand, reduces strain on the grid, and lowers energy costs.

7.3.4 Public Awareness & Education

- **Energy Efficiency Campaigns:** Public awareness campaigns educate consumers about the benefits of energy efficiency and conservation. These campaigns can provide information on energy-saving practices, efficient technologies, and available incentives.
- **School & Community Programs:** Educational programs in schools and communities can teach individuals about energy conservation, sustainable practices, and the importance of reducing energy consumption.

7.3.5 Research & Development

- **Innovation Funding:** Governments and private organizations can fund research and development projects focused on advancing energy-efficient technologies and solutions. This includes support for emerging technologies such as smart grids, energy storage, and advanced materials.
- **Pilot Programs:** Implementing pilot programs allows for testing and demonstration of new energy-efficient technologies and practices. Successful pilots can lead to broader adoption and scaling of effective solutions.

7.3.6 International Agreements & Collaboration

- **Global Initiatives:** International agreements and collaborations, such as the Paris Agreement, set targets and frameworks for reducing greenhouse gas emissions and improving energy efficiency. Participating countries commit to national action plans and share best practices.
- **Cross-Border Projects:** Collaborative projects between countries can promote the exchange of knowledge and technology related to energy

efficiency. These projects may involve joint research, technology transfer, and capacity-building efforts.

7.3.7 Challenges & Future Directions

- **Implementation Barriers:** Effective implementation of energy efficiency policies can face challenges such as lack of funding, limited technical expertise, and resistance to change. Addressing these barriers requires targeted strategies and stakeholder engagement.
- **Monitoring and Evaluation:** Regular monitoring and evaluation of energy efficiency programs are essential to assess their effectiveness and identify areas for improvement. Data-driven approaches can help refine policies and measure progress.
- **Equity Considerations:** Ensuring that energy efficiency programs are accessible to all communities, including low-income and underserved populations, is crucial for achieving equitable benefits. Tailored approaches and support mechanisms can address barriers faced by disadvantaged groups.

As a whole, promoting energy efficiency and conservation through well-designed policies and initiatives is vital for addressing energy challenges and achieving environmental and economic benefits. By implementing standards, providing incentives, raising awareness, and supporting research, governments and organizations can drive meaningful progress towards a more sustainable and efficient energy future.

7.4 References

1. *American Council for an Energy-Efficient Economy (ACEEE). (2023). "Policies for Energy Efficiency". Retrieved from [ACEEE Website](https://www.aceee.org/topics/policies).*
2. *Bioenergy International. (2023). "Biomass and Biofuels". Retrieved from [Bioenergy International](https://www.bioenergyinternational.com/biomass-and-biofuels).*
3. *Environmental Protection Agency (EPA). (2023). "Energy Star Program". Retrieved from [EPA Website](https://www.energystar.gov).*
4. *Geothermal Energy Association (GEA). (2024). "Geothermal Energy Basics". Retrieved from [GEA Website](https://www.geothermal.org/what-geothermal).*
5. *Intergovernmental Panel on Climate Change (IPCC). (2021). "Climate Change 2021: The Physical Science Basis". Cambridge University Press.*
6. *International Energy Agency (IEA). (2023). "Global Energy Review 2023". Retrieved from [IEA Website](https://www.iea.org/reports/global-energy-review-2023).*

7. *International Energy Agency (IEA). (2024). "Energy Efficiency 2024 Report". Retrieved from [IEA Website](https://www.iea.org/reports/energy-efficiency-2024).*
8. *International Hydropower Association (IHA). (2023). "Hydropower Status Report". Retrieved from [IHA Website](https://www.hydropower.org/status-report).*
9. *International Renewable Energy Agency (IRENA). (2023). "Renewable Energy and Energy Efficiency". Retrieved from [IRENA Website](https://www.irena.org/energy-efficiency).*
10. *International Renewable Energy Agency (IRENA). (2023). "Renewable Energy Technologies: Energy for the Future". Retrieved from [IRENA Website](https://www.irena.org/technologies).*
11. *National Renewable Energy Laboratory (NREL). (2024). "Renewable Energy and Energy Efficiency. Retrieved from [NREL Website](https://www.nrel.gov/renewable-energy.html).*
12. *Ocean Energy Council. (2024). "Tidal and Wave Energy". Retrieved from [Ocean Energy Council](https://www.oceanenergycouncil.org).*
13. *U.S. Department of Energy (DOE). (2023). "Energy Efficiency and Renewable Energy". Retrieved from [DOE Website](https://www.energy.gov/eere/energy-efficiency).*
14. *U.S. Department of Energy (DOE). (2024). "Solar Energy Technologies Office". Retrieved from [DOE Website](https://www.energy.gov/eere/solar).*
15. *U.S. Environmental Protection Agency (EPA). (2023). "Air Quality and Emissions". Retrieved from [EPA Website](https://www.epa.gov/air-quality-and-emissions).*

16. *Wind Energy Technologies Office. (2024). "Offshore Wind Research and Development". Retrieved from [DOE Website](https://www.energy.gov/eere/wind/offshore-wind-research-and-development).*
17. *World Nuclear Association. (2024). "The Nuclear Fuel Cycle". Retrieved from [World Nuclear Association](https://www.world-nuclear.org/information-library/current-and-future-generation/the-nuclear-fuel-cycle.aspx).*

Chapter 8: Physics & Environmental Science

Physics plays a critical role in deciphering environmental processes by applying fundamental principles of matter and energy. This chapter explores how physics aids in understanding and predicting environmental phenomena, from climate modelling to ecosystem dynamics.

At its core, physics examines interactions between matter and energy. For instance, nuclear processes such as fission and fusion release significant energy, analogous to atomic interactions in the atmosphere where carbon atoms form various compounds like CO_2. CO_2, a major greenhouse gas, intensifies the greenhouse effect, causing global warming, ice melting, rising sea levels, and extreme weather events. Additionally, carbon compounds can deplete the ozone layer, increasing harmful UV radiation exposure.

Physics principles such as thermodynamics, fluid dynamics, and quantum mechanics are essential in modelling climate systems and predicting changes. These principles enable scientists to create accurate climate models and understand atmospheric interactions.

Advancements in environmental monitoring and modelling technologies, including remote sensing, drones, and high-performance computing, are transforming our ability to track and predict environmental changes. Integrating these technologies with physics principles enhances our understanding of climate change, pollution, and resource management, driving efforts toward sustainable solutions and environmental protection.

8.1 Application of Physics Principles in Understanding Environmental Processes

Physics, fundamentally concerned with the principles of matter and energy, plays a crucial role in understanding various environmental processes. This includes phenomena such as climate modelling and ecosystem dynamics, which are integral to analysing and predicting environmental changes.

At its core, physics examines how matter and energy interact. For instance, the processes of nuclear fission and fusion involve the splitting and combining of atomic nuclei, respectively, which release significant amounts of energy. In environmental contexts, similar principles apply when considering how atoms interact in the atmosphere.

When different atoms are released into the atmosphere, they react with one another to form various molecules. For example, when carbon atoms are emitted, they combine with oxygen atoms to form different compounds such as carbon monoxide (CO), carbon dioxide (CO_2), and carbon tetrachloride (CCl_4). Each of these compounds has distinct environmental impacts.

Carbon dioxide (CO_2) is a significant greenhouse gas that contributes to global warming. Its increased concentration in the atmosphere enhances the greenhouse effect, leading to a rise in Earth's average temperature. This warming has far-reaching consequences, including the melting of polar ice, rising sea levels, and more frequent extreme weather events.

Additionally, the release of carbon-containing compounds can contribute to ozone layer depletion. The ozone layer, located in the stratosphere, absorbs and scatters ultraviolet (UV) radiation from the sun. When this layer is depleted, more UV radiation reaches the Earth's surface, which can lead to adverse effects on human health, wildlife, and plant life.

Understanding these interactions through the lens of physics allows scientists to model climate systems and predict future environmental changes. By applying principles such as thermodynamics, fluid dynamics, and quantum mechanics, researchers can develop more accurate models of how different factors influence climate and ecosystems.

8.2 Technological Advancements in Environmental Monitoring & Modelling

The pursuit of a sustainable and eco-friendly environment is increasingly dependent on advancements in technology, particularly in the realms of environmental monitoring and modelling. These technological innovations are pivotal in our efforts to understand, manage, and mitigate environmental issues such as climate change, pollution, and habitat degradation.

8.2.1 Environmental Monitoring Technologies

Environmental monitoring involves the systematic collection of data related to environmental parameters and pollutants. Advancements in monitoring technologies have significantly enhanced our ability to track and assess various environmental factors with unprecedented accuracy and detail.

- **Remote Sensing:** Remote sensing technology, which uses satellite or aerial sensors to collect data from a distance, has revolutionized environmental monitoring. Satellites such as NASA's Landsat series provide detailed imagery of Earth's surface, enabling scientists to monitor land use changes, deforestation, and urban sprawl. For example, the European Space Agency's Copernicus program offers a suite of Earth observation satellites that track air quality, sea surface temperatures, and vegetation health in near real-time (ESA, 2020).

- **Drones:** Drones equipped with high-resolution cameras and environmental sensors are becoming increasingly popular for monitoring hard-to-reach areas. They offer a flexible and cost-effective way to gather data on wildlife populations, forest health, and pollution levels. Drones can capture detailed images and measurements that are crucial for assessing environmental conditions and detecting changes over time (Koh & Wich, 2012).
- **Internet of Things (IoT):** IoT technology involves embedding sensors in various devices to collect and transmit data over the internet. In environmental monitoring, IoT sensors are used to measure air and water quality, weather conditions, and soil moisture. For instance, smart sensors in urban areas can provide real-time data on air pollution, helping to inform public health decisions and policy (Rouse, 2019).

8.2.2 Advances in Environmental Modelling

Environmental modelling uses mathematical and computational tools to simulate environmental processes and predict future conditions. Recent technological advancements have greatly improved the accuracy and complexity of these models.

- **High-Performance Computing:** Advances in high-performance computing (HPC) have enabled scientists to run complex environmental models with greater speed and accuracy. HPC systems can process large datasets and perform intricate simulations, such as climate models that predict global temperature changes and extreme weather events. The use of supercomputers, like the IBM Summit, allows for more precise climate projections and better understanding of climate dynamics (National Oceanic and Atmospheric Administration, 2021).
- **Machine Learning & Artificial Intelligence (AI):** Machine learning and AI technologies are being increasingly integrated into environmental modelling. These technologies can analyse vast amounts of data to identify patterns and make predictions. For

example, AI algorithms are used to model climate change impacts, forecast air quality, and predict the spread of invasive species. Machine learning techniques enhance the capability to process and interpret complex environmental data, leading to more accurate and actionable insights (Rolnick et al., 2019).

- **Integrated Assessment Models (IAMs):** IAMs combine data from various disciplines to assess the interactions between human activities and environmental systems. These models integrate economic, environmental, and social data to evaluate the effects of policy decisions and management strategies. IAMs are crucial for developing sustainable development scenarios and evaluating the potential impacts of climate policies on both the environment and the economy (van Vuuren et al., 2012).

The rapid advancement of technologies in environmental monitoring and modelling is transforming our ability to understand and address environmental challenges. Remote sensing, drones, IoT, high-performance computing, machine learning, and integrated assessment models all contribute to a more comprehensive understanding of environmental systems. These technological innovations are essential for making informed decisions, developing effective policies, and striving towards a more sustainable and eco-friendly future.

8.3 Contributions of Physics to Sustainable Resource Management & Environmental Protection

Physics, the study of matter and energy and their interactions, is instrumental in advancing sustainable resource management and environmental protection. By applying fundamental principles of physics, we can develop technologies and strategies that enhance the efficiency of resource use, minimize environmental impact, and foster sustainability.

8.3.1 Energy Efficiency & Renewable Resources

One of the primary contributions of physics to sustainable resource management is through the development of energy-efficient technologies and renewable energy sources.

- **Energy Conversion & Efficiency:** The principles of thermodynamics, a branch of physics concerned with heat and energy transfer, are crucial in optimizing energy conversion processes. For example, the efficiency of engines and power plants can be improved by understanding and minimizing energy losses. Technologies such as combined heat and power (CHP) systems utilize waste heat from electricity generation to provide heating, thus improving overall energy efficiency (Becker, 2020).

- **Renewable Energy Technologies:** Physics underpins the design and operation of renewable energy technologies. Photovoltaic cells, which convert sunlight into electricity, rely on the photoelectric effect, where photons excite electrons in a semiconductor material. Advances in materials science and quantum mechanics have led to the development of more efficient and cost-effective solar panels (Green et al., 2017). Similarly, wind turbines harness kinetic energy from wind, and their design and efficiency are optimized through fluid dynamics principles to capture the maximum amount of wind energy (Manwell et al., 2010).

8.3.2 Environmental Monitoring & Pollution Control

Physics also plays a vital role in environmental monitoring and pollution control, providing tools and methods for assessing and mitigating environmental pollution.

- **Remote Sensing and Imaging:** Remote sensing technologies, based on principles of optics and electromagnetic radiation, are used to monitor environmental conditions from a distance. Satellite imagery, for instance, allows for the tracking of deforestation, urban sprawl, and changes in land use. The analysis of spectral signatures in satellite data helps in identifying different types of vegetation and assessing their health (Justice et al., 2019).

  Air and Water Quality Monitoring: Sensors based on physical principles, such as electrochemical sensors and optical sensors, are used to monitor air and water quality. For instance, air quality monitoring involves measuring the concentration of pollutants like particulate matter and gases. These measurements are essential for assessing the impact of pollution on health and ecosystems and for enforcing environmental regulations (Gordon, 2020).

8.3.3 Climate Change Mitigation

Physics contributes significantly to understanding and mitigating climate change through the study of greenhouse gases and climate modelling.

- **Greenhouse Gas Dynamics:** The study of radiative transfer and the greenhouse effect helps in understanding how certain gases trap heat in the atmosphere, leading to global warming. Physics principles are used to measure concentrations of greenhouse gases such as carbon dioxide (CO_2) and methane (CH_4), which are critical for climate change studies and policy-making (IPCC, 2021).
- **Climate Modelling:** Climate models, which are based on complex mathematical equations and physical principles, simulate atmospheric and oceanic processes to predict future climate conditions. These models incorporate data on solar radiation, greenhouse gas concentrations, and ocean currents to forecast changes in temperature, precipitation, and sea levels (Stocker et al., 2013).

8.3.4 Resource Conservation

The principles of physics also aid in resource conservation efforts, helping to manage and preserve natural resources more effectively.

- **Material Science:** Advances in material science, driven by physics, contribute to the development of more durable and sustainable materials. For example, the development of high-strength, lightweight materials reduce the need for raw materials and energy consumption in manufacturing (Ashby, 2011).

- **Water Resources Management:** Physics-based models and simulations are used to manage water resources, such as optimizing water usage in agriculture and predicting the impact of droughts and floods. Hydrological models, which use principles of fluid dynamics, help in managing water supplies and designing effective water conservation strategies (Montanari, 2019).

The application of physics is fundamental to advancing sustainable resource management and environmental protection. From enhancing energy efficiency and developing renewable energy technologies to monitoring environmental quality and mitigating climate change, physics provides the tools and principles necessary for creating a more sustainable future. By leveraging these insights, we can improve our resource management practices, reduce environmental impact, and work towards a more eco-friendly world.

8.4 References

1. *Ashby, M.F. (2011). Materials Selection in Mechanical Design. Butterworth-Heinemann. [Link](https://www.elsevier.com/books/materials-selection-in-mechanical-design/ashby/978-0-08-096702-7)*
2. *Becker, A.C. (2020). Combined Heat and Power: A Review of Technology and Policy. Energy Reports, 6, 352-362. [Link](https://doi.org/10.1016/j.egyr.2020.03.005)*
3. *European Space Agency (ESA). (2020). Copernicus Programme. Retrieved from [ESA](https://www.esa.int/Applications/Observing_the_Earth/Copernicus)*
4. *Gordon, C. (2020). Air Quality Monitoring and Pollution Control. Journal of Environmental Monitoring*, 12(4), 678-685. [Link](https://pubs.rsc.org/en/content/articlelanding/2020/em/d0em90011b)*
5. *Green, M.A., Emery, K., Hishikawa, Y., Warta, W., & Adeniyi, A.O. (2017). The Passivated Emitter and Rear Cell: A Review of the Technology. Progress in Photovoltaics: Research and Applications, 25(7), 668-676. [Link](https://doi.org/10.1002/pip.2902)*
6. *Houghton, J.T., et al. (2001). Climate Change 2001: The Scientific Basis. Cambridge University Press. [Link](https://www.cambridge.org/core/books/abs/climate-change-2001-the-scientific-basis/4E5E58B58A34D2E5E92B8F4A3F86D712F)*
7. *IPCC (2021). Climate Change 2021: The Physical Science Basis. Cambridge University Press.[Link](https://www.ipcc.ch/report/ar6/wg1/)*
8. *IPCC (2021). Climate Change 2021: The Physical Science Basis. Cambridge University Press. [Link](https://www.ipcc.ch/report/ar6/wg1/)*

9. *Justice, C.O., et al. (2019). The Use of Remote Sensing for Monitoring Environmental Changes. Remote Sensing of Environment, 231, 111-128. [Link](https://doi.org/10.1016/j.rse.2019.111266)*
10. *Koh, L.P., & Wich, S.A. (2012). Drones for Biodiversity Conservation and Ecological Research. Trends in Ecology & Evolution, 27(6), 348-355. [Link](https://doi.org/10.1016/j.tree.2012.02.013)*
11. *Manwell, J.F., McGowan, J.G., & Rogers, A.L. (2010). Wind Energy Explained: Theory, Design and Application. Wiley. [Link](https://www.wiley.com/en us/Wind+Energy+Explained%3A+Theory%2C+Design+and+Application-p-9780470015002)*
12. *Montanari, A. (2019). Hydrological Modeling and Water Resources Management. Journal of Hydrology, 568, 9-22. [Link](https://doi.org/10.1016/j.jhydrol.2018.11.016)*
13. *National Aeronautics and Space Administration (NASA). (2020). Ozone Layer and Climate Change. [Link](https://www.nasa.gov/ozone-layer-and-climate-change)*
14. *National Oceanic and Atmospheric Administration (NOAA). (2021). Supercomputers and Climate Modeling. Retrieved from [NOAA](https://www.noaa.gov/supercomputers-climate-modeling)*
15. *Rolnick, D., et al. (2019). Tackling Climate Change with Machine Learning*. ACM Computing Surveys, 52(3), 1-35. [Link](https://doi.org/10.1145/3293663)*
16. *Rouse, J. (2019). The Role of IoT in Environmental Monitoring. IoT Journal. Retrieved from [IoT Journal](https://www.iotjournal.com/role-iot-environmental-monitoring)*

17. *Seinfeld, J.H., & Pandis, S.N. (2016). Atmospheric Chemistry and Physics: From Air Pollution to Climate Change. Wiley. [Link](https://www.wiley.com/en us/Atmospheric+Chemistry+and+Physics%3A+From+Air+Pollution+to+Climate+Change%2C+3rd+Edition-p-9781119475178)*
18. *Stocker, T.F., et al. (2013). Climate Change 2013: The Physical Science Basis. Cambridge University Press. [Link](https://www.cambridge.org/core/books/climate-change-2013-the-physical-science-basis/CE3D9F825C9E5874F1D36C885AB185A4)*
19. *van Vuuren, D.P., et al. (2012). The Representation of Energy Technologies in Integrated Assessment Models. Energy Policy, 41, 219-237. [Link](https://doi.org/10.1016/j.enpol.2011.10.007)*

Chapter 9: Environmental Policy & Governance

International organizations are pivotal in shaping global environmental agendas and fostering cooperative actions to address pressing environmental challenges. Key institutions like the United Nations (UN) and the Intergovernmental Panel on Climate Change (IPCC) lead these efforts by providing frameworks for international cooperation, scientific assessment, and policy development.

The UN, through its various bodies such as the UN Environment Programme (UNEP), offers a central platform for global environmental diplomacy and policy-making. UNEP coordinates international environmental activities, develops agreements, and promotes sustainable development. The UN's Sustainable Development Goals (SDGs), particularly Goal 13 on climate action, provide a comprehensive framework for integrating environmental sustainability into national policies.

The IPCC, established by the World Meteorological Organization and UNEP, serves as the foremost authority on climate change science. Its comprehensive assessment reports and special reports offer critical insights into climate impacts and guide international climate policy, including landmark agreements like the Paris Agreement.

These organizations also drive collaborative efforts and future directions, such as the UN's "Decade on Ecosystem Restoration" and the IPCC's on-going research. By fostering international cooperation and integrating scientific research into policy, they are crucial in advancing global environmental agendas and ensuring a sustainable future.

9.1 Role of International Organizations in Shaping Global Environmental Agendas

International organizations play a crucial role in shaping global environmental agendas, guiding international cooperation, and fostering policies aimed at addressing complex environmental challenges. Institutions such as the United Nations (UN) and the Intergovernmental Panel on Climate Change (IPCC) are central to these efforts, providing frameworks for collaborative action, scientific assessment, and policy formulation. On the other need International organisations to need to pay more attention to those regions of the world which are being severely affected due to climate changes.

9.1.1 The United Nations: A Platform for Global Cooperation

The United Nations (UN) serves as a central platform for international environmental diplomacy and policy-making. Established in 1945, the UN has evolved to address a wide range of global issues, including environmental sustainability.

- **UN Environment Programme (UNEP):** Founded in 1972, UNEP is a leading global environmental authority within the UN system. It coordinates environmental activities, develops international environmental agreements, and promotes sustainable development through sound environmental practices. UNEP's reports and assessments, such as the "Global Environment Outlook" series, provide critical data and analysis on global environmental trends and issues (UNEP, 2019).
- **Sustainable Development Goals (SDGs):** The 2030 Agenda for Sustainable Development, adopted by all UN Member States in 2015, includes 17 Sustainable Development Goals (SDGs) that address various aspects of global sustainability, including environmental protection. Goal 13 specifically focuses on climate action, while

Goals 6, 14, and 15 address to water, ocean, and land conservation, respectively. The SDGs offer a comprehensive framework for integrating environmental sustainability into national policies and development plans (United Nations, 2020).

- **UN Framework Convention on Climate Change (UNFCCC):** The UNFCCC is a critical international treaty aimed at addressing climate change. Established in 1992, it serves as the foundation for international climate negotiations and agreements. The UNFCCC's Conference of the Parties (COP) meetings are key events where countries negotiate and update their climate commitments. The Paris Agreement, adopted at COP21 in 2015, represents a landmark achievement in global climate diplomacy, setting targets for limiting global warming and enhancing climate resilience (UNFCCC, 2015).

9.1.2 Intergovernmental Panel on Climate Change: A Scientific Authority

The IPCC, established in 1988 by the World Meteorological Organization (WMO) and UNEP, is the leading international body for the assessment of climate change. The IPCC's primary role is to provide policymakers with regular scientific assessments on climate change, its impacts, and potential adaptation and mitigation strategies.

- **Assessment Reports:** The IPCC produces comprehensive assessment reports, which are widely regarded as the most authoritative source of scientific knowledge on climate change. These reports synthesize the latest research and provide a detailed understanding of the state of the climate system, projected future changes, and policy options. The latest assessment report, "Climate Change 2021: The Physical Science Basis," outlines the accelerating pace of climate change and the urgent need for action (IPCC, 2021).
- **Special Reports:** In addition to its main assessment reports, the IPCC also publishes special reports on specific topics, such as the impacts of global warming of 1.5°C above pre-industrial levels and the implications for ecosystems and human societies. These reports provide targeted insights that help guide international climate policy and inform the public and stakeholders about critical issues (IPCC, 2018).
- **Role in Policy Formulation:** The IPCC's scientific assessments are instrumental in shaping global climate policies. By providing a

rigorous scientific basis for understanding climate change, the IPCC influences negotiations and decisions made by international bodies, national governments, and other stakeholders. Its findings are often used to support the development of climate action plans, adaptation strategies, and mitigation measures.

9.1.3 Collaborative Efforts & Future Directions

The collaboration between the UN, IPCC, and other international organizations underscores the collective effort required to tackle global environmental challenges. Initiatives such as the UN's "Decade on Ecosystem Restoration" and the IPCC's on-going research into climate change impacts demonstrate the commitment to advancing environmental protection and sustainability.

Looking ahead, it is essential for these organizations to continue fostering international cooperation, addressing emerging environmental issues, and integrating scientific research into policy frameworks. Strengthening the collaboration between governments, scientists, and civil society will be crucial in achieving the global environmental agendas and ensuring a sustainable future for all.

9.2 Influence of Political Leaders & Policymakers in Driving Environmental Initiatives

Political leaders and policymakers play a pivotal role in shaping environmental initiatives and driving progress toward sustainability. Their influence extends from setting regulatory frameworks and international agreements to mobilizing public support and funding for environmental projects. By leveraging their authority and platforms, they can significantly impact environmental protection and sustainability efforts on both national and global scales.

9.2.1 Setting the Agenda: Policy Formulation & Legislation

One of the primary ways political leaders and policymakers influence environmental initiatives is through the formulation and enactment of policies and legislation. These decisions establish the regulatory frameworks that govern environmental practices and standards.

- **Environmental Legislation:** National and local governments create laws that regulate environmental issues such as air and water quality, waste management, and conservation. For example, the Clean Air Act and the Clean Water Act in the United States set stringent standards for pollution control and have been instrumental in improving environmental quality since their enactment (EPA, 2020). Similarly, the European Union's Environmental Directives establish comprehensive regulations for managing various environmental aspects, including waste, water, and air pollution (European Commission, 2021).

- **Climate Policies:** Political leaders play a crucial role in shaping climate policies, which are essential for addressing global warming and mitigating climate change impacts. The Paris Agreement, adopted in 2015, is a prime example of how international negotiations and commitments, spearheaded by political leaders, aim to limit global temperature rise and enhance climate resilience (UNFCCC, 2015). Policymakers are responsible for setting national targets for greenhouse gas reductions and implementing strategies to meet these commitments.

9.2.2 Mobilizing Resources & Funding

Political leaders and policymakers also influence environmental initiatives by mobilizing resources and securing funding for environmental projects and research.

- **Government Funding:** Investment in environmental initiatives often requires substantial financial resources. Political leaders can allocate government funds to support research, development, and implementation of environmental technologies and programs. For instance, funding for renewable energy projects, such as wind and solar power, is often provided through government subsidies and incentives. These financial resources are crucial for advancing clean energy technologies and reducing dependence on fossil fuels (International Energy Agency, 2020).
- **Public & Private Sector Collaboration:** Leaders can foster partnerships between the public and private sectors to address environmental challenges. By creating favourable policy environments and providing incentives, policymakers encourage private companies to invest in sustainable practices and technologies.

Public-private partnerships can accelerate innovation and deployment of environmental solutions, such as green infrastructure and sustainable agriculture practices (World Bank, 2019).

9.2.3 Leading International Cooperation

On the global stage, political leaders and policymakers are essential in driving international cooperation on various environmental issues.

- **Global Agreements & Conventions:** International agreements and conventions, such as the Kyoto Protocol and the Paris Agreement, are negotiated and ratified by political leaders from various countries. These agreements set collective goals and commitments for addressing global environmental issues, such as climate change and biodiversity loss. The success of these agreements depends on the commitment and leadership of policymakers who champion these causes at international forums (UNFCCC, 2015).
- **Advocacy & Diplomacy:** Political leaders use their influence to advocate for environmental issues and build consensus among nations. High-profile leaders can draw attention to pressing environmental challenges and rally support for global initiatives. For

example, former U.S. President Al Gore's advocacy on climate change has significantly raised public awareness and influenced policy discussions globally (Gore, 2006).

9.2.4 Driving Public Engagement & Awareness

Political leaders and policymakers also play a role in driving public engagement and raising awareness about environmental issues.

- **Public Campaigns & Education:** Leaders can initiate public campaigns and educational programs to inform citizens about environmental issues and encourage sustainable behaviours. By addressing environmental challenges in speeches, media appearances, and policy platforms, political figures can elevate these issues in the public consciousness and foster greater community involvement in environmental protection (National Geographic Society, 2021).
- **Community Leadership:** Local leaders and policymakers influence environmental initiatives by addressing community-specific issues and promoting local sustainability projects. Their involvement in grassroots efforts and community-based environmental programs can

drive positive change at the local level and serve as models for broader adoption.

Political leaders and policymakers are instrumental in advancing environmental initiatives through policy formulation, resource mobilization, international cooperation, and public engagement. Their decisions and actions shape the trajectory of environmental protection efforts and influence the effectiveness of sustainability strategies. By harnessing their leadership and authority, they can drive meaningful progress toward a more sustainable and resilient future.

9.3 Case Studies of Successful Environmental Policies & Their Impact on Sustainability

In the 21st century, the global shift from non-renewable to renewable energy sources marks a transformative period in environmental policy. This shift is driven by the urgent need to combat climate change and reduce carbon emissions, which have been exacerbated by the extensive use of fossil fuels. Governments, organizations, and communities worldwide are increasingly adopting policies that promote renewable energy, such as solar and wind power. These policies not only mitigate the environmental impact of traditional energy sources but also pave the way for a sustainable future.

Several case studies illustrate how successful environmental policies can drive a significant change. One notable example is Germany's Energiewende, or "energy transition." Since the early 2000s, Germany has implemented a comprehensive policy framework aimed at transitioning from fossil fuels to renewable energy. The Energiewende encompasses ambitious targets for reducing greenhouse gas emissions, increasing the share of renewable energy in the energy mix, and improving energy efficiency. The policy includes incentives for the installation of solar panels and wind turbines, as well as subsidies for energy-efficient home renovations.

The impact of the Energiewende has been profound. By 2023, Germany had achieved around 50% of its electricity consumption from renewable sources, up from just 6% in 2000. The transition has also contributed to significant reductions in carbon emissions, with Germany cutting its emissions by approximately 40% from 1990 levels by 2023. Moreover, the Energiewende has stimulated job creation in the renewable energy sector, providing employment for over 300,000 individuals.

Another compelling case study is that of Denmark, which has set a global benchmark for the wind energy utilization. Denmark's commitment to renewable energy began in the 1970s with a focus on developing wind power as a response to the oil crises. The Danish government introduced policies that supported research and development in wind technology, provided financial incentives for wind farm construction, and established a stable regulatory environment for investors.

By 2023, wind energy accounted for over 50% of Denmark's total electricity consumption. The country has become a leader in wind turbine technology and exports its expertise worldwide. Denmark's success in wind energy has not only reduced its carbon footprint but also enhanced its energy security and economic growth. The country has demonstrated that a strategic focus on renewable energy can lead to both environmental and economic benefits.

In the United States, California serves as a prominent example of state-level environmental policy. California's Renewable Portfolio Standard (RPS), enacted in 2002, mandates that a significant percentage of the state's electricity must come from renewable sources. The policy has been progressively strengthened over the years, with the state setting a goal of 60% renewable energy by 2030 and 100% carbon-free electricity by 2045.

The RPS has driven substantial investment in solar and wind energy infrastructure in California. The state now leads the nation in solar power capacity, with over 30,000 megawatts installed. California's shift towards renewable energy has contributed to a notable decrease in greenhouse gas emissions and has positioned the state as a global leader in clean energy innovation.

These case studies highlight the potential for environmental policies to drive substantial progress in sustainability. By setting ambitious targets, providing financial incentives, and fostering innovation, countries and states can transition away from fossil fuels and embrace renewable energy sources. The experiences of Germany, Denmark, and California underscore the importance of comprehensive policy frameworks in achieving significant environmental and economic outcomes.

As the world continues to grapple with climate change, the lessons learned from these successful case studies can serve as a guide for other regions looking to implement their own sustainable energy policies. The transition to renewable energy is not only a crucial step in reducing carbon emissions but also a vital component of building a resilient and sustainable future for generations to come.

9.4 References

1. *California Renewable Portfolio Standard (RPS). California Public Utilities Commission. https://www.cpuc.ca.gov/renewable-portfolio-standard*
2. *Denmark's Wind Energy Journey. Danish Energy Agency. https://ens.dk/en/our-services/wind-energy*
3. *Energiewende: Germany's Energy Transition. Federal Ministry for Economic Affairs and Energy. https://www.bmwi.de/Redaktion/EN/Artikel/Energy/energiewende.html*
4. *Environmental Protection Agency (EPA). (2020). Clean Air Act Overview. Retrieved from [EPA](https://www.epa.gov/clean-air-act-overview)*
5. *European Commission. (2021). Environmental Directives. Retrieved from [European Commission](https://ec.europa.eu/environment/)*
6. *Gore, A. (2006). An Inconvenient Truth: The Planetary Emergency of Global Warming and What We Can Do About It. Rodale Books.*
7. *Højstrup, J., & Bering, J. (2014). "Denmark: An international leader in wind energy." Energy, 72, 92-100.*
8. *Intergovernmental Panel on Climate Change (IPCC). (2018). Global Warming of 1.5°C. Retrieved from [IPCC](https://www.ipcc.ch/sr15/)*

9. *Intergovernmental Panel on Climate Change (IPCC). (2021). Climate Change 2021: The Physical Science Basis. Cambridge University Press. Retrieved from [IPCC](https://www.ipcc.ch/report/ar6/wg1/)*
10. *International Energy Agency (IEA). (2020). World Energy Investment 2020. Retrieved from [IEA](https://www.iea.org/reports/world-energy-investment-2020)*
11. *Jacobson, M. Z., & Delucchi, M. A. (2011). "Providing all global energy with wind, water, and solar power." Energy Policy, 39(3), 1154-1169.*
12. *Muro, M., & Saha, D. (2015). "California's clean energy future: The case for a climate and energy policy agenda." Brookings Institution. [https://www.brookings.edu/research/californias-clean-energy future](https://www.brookings.edu/research/californias-clean-energy-future)*
13. *National Geographic Society. (2021). Environment and Sustainability. Retrieved from [National Geographic](https://www.nationalgeographic.com/environment/)*
14. *UN Framework Convention on Climate Change (UNFCCC). (2015). Paris Agreement. Retrieved from [UNFCCC](https://unfccc.int/process-and-meetings/the-paris-agreement/the-paris-agreement)*
15. *UN Framework Convention on Climate Change (UNFCCC). (2015). Paris Agreement. Retrieved from [UNFCCC](https://unfccc.int/process-and-meetings/the-paris-agreement/the-paris-agreement)*
16. *United Nations (UN). (2020). The 2030 Agenda for Sustainable Development. Retrieved from [United Nations](https://sdgs.un.org/2030agenda)*
17. *United Nations Environment Programme (UNEP). (2019). Global Environment Outlook – GEO-6: Healthy Planet, Healthy People. Retrieved*

from [UNEP](https://www.unep.org/resources/report/global-environment-outlook-6)

18. *World Bank. (2019). Public-Private Partnerships for Sustainable Development. Retrieved from [World Bank](https://www.worldbank.org/en/topic/publicprivatepartnerships)*

Chapter 10: Circular Economy: Redefining Resource Management

The circular economy represents a transformative shift from traditional linear models of resource use, emphasizing a closed-loop system where products are designed to be reused, repaired, remanufactured, and recycled. This approach not only reduces waste and minimizes environmental impact but also fosters economic resilience and innovation. Key principles of the circular economy include designing for longevity, optimizing resource efficiency, adopting product-as-a-service models, and implementing closed-loop systems. These principles involve diverse stakeholders, designers, engineers, entrepreneurs, consumers, material scientists, product managers, and quality control inspectors, each playing a crucial role in ensuring products are sustainable throughout their lifecycle.

The impact of circular economy practices extends to sustainable development by aligning with global sustainability goals, including the United Nations Sustainable Development Goals (SDGs). Case studies, such as Ecolab's closed-loop water system and Patagonia's Worn Wear program, illustrate successful circular practices in waste management and resource recovery. However, scaling these practices globally faces challenges such as infrastructure limitations, financial barriers, regulatory constraints, and cultural shifts. Despite these obstacles, opportunities exist in technological innovation, economic benefits, consumer demand, policy support, and global collaboration, offering pathways to a more sustainable and resilient global economy.

10.1 Circular Economy & Its Role in Sustainable Development

The circular economy is a transformative approach to economic and environmental sustainability that redefines how products are designed, produced, consumed, and disposed of. Unlike the traditional linear economy, where resources are extracted, used, and then discarded, the circular economy aims to create a closed-loop system where resources are continually cycled back into the production process.

This model not only seeks to minimize waste and reduce environmental impact but also aims to foster economic resilience and innovation. Understanding the principles of the circular economy and its role in sustainable development involves examining the interconnected roles of various stakeholders, including designers, engineers, entrepreneurs, consumers, material scientists, product managers, and quality control inspectors.

10.1.1 Key Principles of the Circular Economy

- **Design for Longevity:** The principle of designing for longevity involves creating products that are durable, easy to repair, and upgradeable. By focusing on quality and longevity, manufacturers can extend the life cycle of products and reduce the frequency of replacements. For instance, modular design allows for individual components to be replaced or upgraded without discarding the entire product. This approach not only conserves resources but also reduces waste.
- **Resource Efficiency:** Resource efficiency aims to maximize the utility of materials and energy throughout the product life cycle. This involves optimizing the use of raw materials, reducing energy consumption during production, and minimizing waste. Techniques such as recycling, remanufacturing, and up-cycling are central to this principle. By improving resource efficiency, businesses can lower costs and reduce their environmental footprint.

- **Product-as-a-Service:** This model shifts the focus from ownership to usage. Instead of selling products, companies provide services where products are leased or rented. This approach encourages manufacturers to design products that are durable and easy to maintain since they retain ownership and responsibility for the product's lifecycle. For example, companies like Philips offer lighting-as-a-service, where clients pay for the use of lighting systems rather than purchasing the fixtures outright.
- **Closed-Loop Systems:** Closed-loop systems aim to create a cycle where waste materials are continuously reused and recycled. This involves designing products that can be easily disassembled and materials that can be reclaimed and repurposed. For example, in the fashion industry, brands like Patagonia implement closed-loop systems by encouraging customers to return worn items for recycling or refurbishment.
- **Collaboration & Systems Thinking:** The circular economy relies on the collaboration of various stakeholders across the value chain. This involves engaging designers, engineers, entrepreneurs, material scientists, product managers, and quality control inspectors in a

systems-based approach. By working together, these stakeholders can ensure that products are designed with the entire lifecycle in mind, from raw material extraction to end-of-life disposal.

10.1.2 Role of Stakeholders in Circular Economy

- **Designers:** Designers play a crucial role in the circular economy by creating products that are not only aesthetically pleasing but also functional and sustainable. They must consider factors such as material choice, ease of disassembly, and repair-ability. Designers who embrace circular principles contribute to the development of products that can be easily recycled or up-cycled, reducing waste and resource consumption.
- **Engineers:** Engineers are responsible for translating design concepts into practical, functional products. In the circular economy, engineers focus on developing technologies and processes that enhance resource efficiency and product longevity. This includes innovations in manufacturing techniques, materials science, and energy-efficient technologies. Engineers also play a role in designing systems for product recovery and recycling.

- **Entrepreneurs:** Entrepreneurs drive the adoption of circular economy principles by developing new business models and innovative solutions. They identify opportunities to create value through circular practices, such as product-as-a-service models or take-back schemes. Entrepreneurs are also instrumental in fostering a culture of sustainability and encouraging other businesses to adopt circular practices.
- **Consumers:** Consumers are essential participants in the circular economy. Their choices and behaviours influence demand for sustainable products and services. By opting for products that are designed for durability, repair-ability, and recyclability, consumers can support the transition to a circular economy. Educating consumers about the benefits of circular practices and encouraging responsible consumption are the key to driving change.
- **Material Scientists:** Material scientists contribute to the circular economy by developing new materials that are sustainable, recyclable, and less harmful to the environment. They focus on creating materials that can be easily recovered and reused, thereby reducing the reliance

on virgin resources. Advances in materials science play a vital role in enabling closed-loop systems and improving resource efficiency.

- **Product Managers:** Product managers oversee the lifecycle of products from conception to end-of-life. They are responsible for ensuring that products meet customer needs while adhering to circular economy principles. This includes coordinating with designers and engineers to incorporate sustainability features, managing supply chains to minimize waste, and implementing take-back programs for end-of-life products.
- **Quality Control Inspectors:** Quality control inspectors ensure that products meet performance and safety standards throughout their lifecycle. In the circular economy, their role extends to verifying that products are designed for longevity and can be effectively repaired or recycled. Quality control is crucial for maintaining consumer trust and ensuring that circular practices are implemented effectively.

10.1.3 Impact on Sustainable Development

The adoption of circular economy principles has significant implications for sustainable development. By minimizing waste and optimizing resource use, the circular economy contributes to environmental protection and reduces the strain on natural resources. This approach also promotes economic resilience by creating new business opportunities and driving innovation.

Furthermore, the circular economy supports social sustainability by fostering a culture of responsible consumption and production. It encourages businesses to engage with communities and educate consumers about the benefits of circular practices. This holistic approach aligns with global sustainability goals, including those outlined in the United Nations Sustainable Development Goals (SDGs).

In summary, the circular economy represents a paradigm shift in how we approach resource management and sustainability. By integrating the principles of design for longevity, resource efficiency, product-as-a-service, closed-loop systems, and collaboration, stakeholders across various sectors can contribute to a more sustainable and resilient future. The collective efforts of designers, engineers, entrepreneurs, consumers, material scientists, product managers, and quality control inspectors are essential in advancing the circular economy and achieving sustainable development goals.

10.2 Case Studies of Circular Economy Practices in Industries

The circular economy represents a significant shift from the traditional linear model of "take, make, dispose." Instead of this one-way trajectory, the circular economy seeks to create a closed-loop system where resources are continually reused, repaired, remanufactured, and recycled. This paradigm shift is crucial in addressing the environmental challenges posed by conventional manufacturing and consumption patterns.

The insights shared by designer Sophie Thomas during her TEDx Talk in Berlin on May 7, 2024, highlight the pressing need for a transformation in design and manufacturing processes. Thomas emphasized that many modern products are designed with little regard for their environmental impact, leading to significant waste and pollution. Her observations underscore the urgency of adopting circular economy principles across various industries, including waste management and resource recovery.

10.2.1 Key Issues Highlighted by Sophie Thomas

Sophie Thomas's TEDx Talk sheds light on several critical issues related to product designing and its environmental impact. She pointed out that many products are designed without considering their end-of-life stage, resulting in substantial waste. For instance, the production of a single toothbrush involves 1.5 kilograms of material, much of which ends up as waste. Similarly, two tons of raw materials are required to produce a 5-gram gold ring, and the creation of a pair of shoes consumes approximately 8,000 liters of water. Alarmingly, about 90% of raw materials become waste before products even reach the market, and 80% of products are discarded within six months of use. This wasteful process not only strains natural resources but also contributes to pollution, especially in more advanced and industrialized societies.

10.2.2 Circular Economy Practices in Waste Management

- **Ecolab's Closed-Loop Water System:** Ecolab, a global leader in water, hygiene, and energy technologies, has implemented a closed-loop water system to address industrial water waste. In their facilities, Ecolab has developed a system that recycles water from manufacturing processes and reuses it, significantly reducing the need for fresh water and minimizing wastewater generation. This approach aligns with circular economy principles by ensuring that water resources are continually cycled through the system rather than being discarded.

- **Waste Management in the City of Ljubljana:** Ljubljana, the capital of Slovenia, has become a model for circular waste management. The city has implemented a comprehensive waste separation and recycling program, achieving a 68% recycling rate by 2023. Ljubljana's approach includes extensive public education campaigns, the introduction of separate waste bins for different materials, and partnerships with local businesses to promote waste reduction. The city's success demonstrates how circular economy practices can be effectively applied at the municipal level to reduce landfill waste and enhance resource recovery.

10.2.3 Circular Economy Practices in Resource Recovery

- **Patagonia's Worn Wear Program:** Patagonia, the outdoor apparel company, has pioneered the circular economy in the fashion industry through its Worn Wear program. This initiative encourages customers to repair, reuse, and recycle their Patagonia products. The company offers repair services for worn items, facilitates the resale of used clothing, and recycles materials from end-of-life products into new garments. Patagonia's approach not only extends the life of its products but also reduces the environmental impact associated with garment production and disposal.

- **Interface's Re-entry Program:** Interface, a global manufacturer of modular carpets, has implemented the Re-entry program to promote circularity in carpet production. The program focuses on recovering and recycling used carpet tiles, which are then reprocessed into new products. Interface's efforts have led to the diversion of millions of pounds of carpet waste from landfills and the reduction of raw material consumption. By integrating circular economy principles into its business model, Interface demonstrates how resource recovery can be effectively incorporated into manufacturing processes.

10.2.4 Role of Designers in Promoting Circular Economy

Designers play a pivotal role in facilitating the transition to a circular economy. Their choices and practices significantly impact the sustainability of products and systems. According to Thomas, designers must shift their focus from creating products with planned obsolescence to designing for durability, repair-ability, and recyclability. This involves:

- **Designing for Longevity:** Products should be designed to last longer and withstand wear and tear. This includes using durable materials and modular designs that allow for easy repairs and upgrades.
- **Emphasizing Material Flow:** Designers should consider the entire lifecycle of a product, including its end-of-life stage. This involves selecting materials that can be easily recycled or repurposed and designing products that can be disassembled for recycling.
- **Inspiring Behavioural Change:** Designers can influence consumer behaviour by creating products that encourage sustainable practices. For example, designing products that are easy to repair or refurbish can promote a culture of maintenance rather than disposal.
- **Communicating Environmental Impact:** Effective communication of a product's environmental impact can raise awareness among consumers and drive demand for more sustainable options. Designers should provide clear information about the sustainability features of their products and the benefits of circular practices.

The transition to a circular economy represents a crucial step in addressing the environmental challenges posed by traditional linear models of production and consumption. The case studies of Ecolab's closed-loop water system, Ljubljana's waste management program, Patagonia's Worn Wear initiative, and Interface's Re-entry program illustrate the potential for circular economy practices to drive meaningful change in waste management and resource recovery. Sophie Thomas's TEDx Talk highlights the urgent need for designers to embrace circular principles and consider the environmental impact of their creations. By focusing on durability, repair-ability, and material flow, and by inspiring behavioural changes and effective communication, designers can play a key role in advancing the circular economy and fostering a more sustainable future.

10.3 Challenges & Opportunities for Scaling Up Circular Economy Models Globally

The concept of a circular economy (CE) is gaining momentum as a viable alternative to the traditional linear model of "take, make, dispose." The circular economy promotes a regenerative system where resources are continuously cycled through reuse, repair, remanufacturing, and recycling.

This model not only aims to minimize waste and reduce environmental impact but also to create economic value and foster sustainable development. However, while the potential benefits of circular economy models are substantial, scaling these practices globally presents a range of challenges and opportunities. Addressing these challenges and seizing the opportunities will be crucial for advancing the circular economy on a global scale.

10.3.1 Challenges in Scaling Up Circular Economy Models

- **Infrastructure & Technological Limitations:** One of the significant challenges in scaling up circular economy models is the lack of adequate infrastructure and technology. Efficient recycling and resource recovery systems require sophisticated facilities and advanced technologies. In many regions, especially in developing countries, such infrastructure is either non-existent or underdeveloped. This limitation hampers the ability to effectively close the loop in material cycles and recycle products back into the supply chain. Additionally, technology for material recovery and product refurbishment needs continuous innovation to handle diverse materials and complex products.

- **Economic & Financial Barriers:** Transitioning to a circular economy often involves significant upfront investment. Businesses may face high costs associated with redesigning products for durability, establishing new recycling systems, or adopting new technologies. For small and medium-sized enterprises (SMEs), these costs can be prohibitive. Furthermore, existing financial models and incentives are frequently geared towards linear processes, making it challenging for circular economy initiatives to secure funding and support. The financial viability of circular economy projects must be demonstrated through clear economic benefits and returns on investment.
- **Regulatory & Policy Constraints:** The regulatory landscape can both support and hinder the transition to a circular economy. In many regions, policies and regulations are still rooted in the linear economic model. This includes standards and regulations that prioritize new product manufacturing over recycling or remanufacturing. The absence of comprehensive circular economy policies and incentives can stifle innovation and slow down the adoption of circular practices. Governments and policymakers need to develop and implement

regulations that facilitate the transition to a circular economy, including extended producer responsibility, waste reduction targets, and recycling mandates.

- **Cultural & Behavioural Challenges:** Shifting from a linear to a circular economy also requires a significant cultural and behavioural change. Consumers and businesses must adopt new practices such as repair, reuse, and recycling. In many cultures, there is a strong preference for new overused products, and the convenience of disposability often outweighs the perceived benefits of circular practices. Changing consumer behaviour and attitudes towards product ownership and waste is essential for the success of circular economy models. This requires effective education and awareness campaigns to highlight the benefits of circular practices and foster a culture of sustainability.
- **Supply Chain Complexity:** The complexity of global supply chains poses another challenge for scaling circular economy models. Products are often made from a mix of materials sourced from different parts of the world, making it difficult to track and manage their end-of-life stages. Additionally, supply chains for recycling and

resource recovery are often fragmented and inefficient. Coordinating efforts across diverse stakeholders and managing the flow of materials through complex supply chains require enhanced collaboration and transparency. Developing effective reverse logistics systems and ensuring the traceability of materials are crucial for achieving circularity.

10.3.2 Opportunities for Scaling Up Circular Economy Models

- **Innovation & Technology Development:** The push towards a circular economy creates opportunities for innovation and technological advancement. New technologies and business models are emerging that facilitate product repair, material recovery, and waste reduction. Innovations such as advanced recycling technologies, digital tracking systems for materials, and sustainable product design are paving the way for circular practices. Investing in research and development can accelerate the adoption of these technologies and make circular economy models more accessible and effective.
- **Economic Benefits & Job Creation:** Adopting circular economy practices can yield significant economic benefits. Circular models can

reduce costs associated with raw material procurement, waste management, and energy consumption. Moreover, the circular economy has the potential to create new business opportunities and jobs. For example, sectors such as repair services, recycling industries, and product refurbishment can thrive in a circular economy. The development of circular supply chains and closed-loop systems can stimulate economic growth and foster local job creation.

- **Consumer Demand & Market Opportunities:** There is a growing consumer demand for sustainable products and practices. Consumers are increasingly aware of environmental issues and are seeking products that align with their values. Businesses that adopt circular economy principles can differentiate themselves in the market by offering sustainable products and services. This growing market demand presents opportunities for companies to build brand loyalty, capture new customer segments, and drive innovation. By meeting consumer expectations for sustainability, businesses can enhance their competitive advantage.
- **Policy & Regulatory Support:** Governments and international organizations are increasingly recognizing the importance of the

circular economy and are beginning to develop supportive policies and regulations. The European Union, for instance, has established a Circular Economy Action Plan that outlines measures to support circular practices across various sectors. Other countries and regions are also implementing policies that promote recycling, waste reduction, and sustainable product design. Leveraging these policy developments can create a favourable environment for scaling circular economy models and provide businesses with the necessary incentives and support.

- **Global Collaboration & Knowledge Sharing:** Scaling up circular economy models globally requires collaboration and knowledge sharing among stakeholders. International organizations, industry groups, and academic institutions can play a crucial role in facilitating the exchange of best practices and innovative solutions. Platforms for collaboration and knowledge sharing can help address common challenges, disseminate successful case studies, and build capacity for circular practices. By working together, stakeholders can accelerate the adoption of circular economy models and overcome barriers to implementation.

Scaling up circular economy models globally presents both significant challenges and promising opportunities. Addressing challenges such as infrastructure limitations, economic barriers, regulatory constraints, cultural shifts, and supply chain complexities is essential for advancing circular practices. At the same time, embracing opportunities related to innovation, economic benefits, consumer demand, policy support, and global collaboration can drive the successful implementation of circular economy models. By navigating these challenges and leveraging opportunities, stakeholders can contribute to a more sustainable and resilient global economy that aligns with the principles of circularity.

10.4 References

1. *Bocken, N. M. P., de Pauw, I., Bakker, C., & van der Grinten, B. (2016). "Product design and business model strategies for a circular economy." Journal of Cleaner Production, 97, 45-56.*
2. *Circular Economy: A New Paradigm for Sustainable Development. World Economic Forum. https://www.weforum.org/reports/circular-economy*
3. *City of Ljubljana. "Waste Management." https://www.ljubljana.si/en/environment-and-sustainable-development/waste-management/*

4. *Dauriat, C., & Smith, J. (2020). "Systems Thinking in Circular Economy: A Comprehensive Review." Systems, 8(4), 38-47.*
5. *Ecolab. "Water Reuse and Recycling." https://en-us.ecolab.com/solutions/water-reuse-and-recycling*
6. *Ellen MacArthur Foundation. (2015). "Growth Within: A Circular Economy Vision for a Competitive Europe." https://www.ellenmacarthurfoundation.org/assets/downloads/publications/EllenMacArthurFoundation_GrowthWithin_July2015.pdf*
7. *Ellen MacArthur Foundation. (2021). "The Circular Economy Opportunity." https://www.ellenmacarthurfoundation.org/circular-economy*
8. *Ellen MacArthur Foundation. "Designing for Longevity." https://www.ellenmacarthurfoundation.org/assets/downloads/Designing-for-Longevity.pdf*
9. *European Commission. (2020). "Circular Economy Action Plan for a Cleaner and More Competitive Europe." https://ec.europa.eu/environment/circular-economy/index_en.htm*
10. *Geissdoerfer, M., et al. (2018). "The Circular Economy – A new sustainability paradigm?" *Journal of Cleaner Production*, 143, 757-768.*

https://doi.org/10.1016/j.jclepro.2016.12.048

11. *Interface. "ReEntry 2.0: Carpet Recycling and Reuse." [https://www.interface.com/APAC/en AU/sustainability/reentry](https://www.interface.com/APAC/en-AU/sustainability/reentry)*
12. *Kirchherr, J., Reike, D., & Hekkert, M. (2017). "Conceptualizing the circular economy: An analysis of 114 definitions." Resources, Conservation and Recycling, 127, 221-232.*
13. *Lacy, P., & Rutqvist, J. (2016). "Waste to Wealth: The Circular Economy Advantage." Palgrave Macmillan.*
14. *Lacy, P., & Rutqvist, J. (2016). "Waste to Wealth: The Circular Economy Advantage." Palgrave Macmillan.*
15. *McDonough, W., & Braungart, M. (2002). "Cradle to Cradle: Remaking the Way We Make Things." North Point Press.*
16. *Mont, O., & Plepys, A. (2008). "Sustainability and Consumer Behavior: A Review." *Journal of Cleaner Production*, 16(4), 373-378. https://doi.org/10.1016/j.jclepro.2007.01.008*
17. *Nielsen. (2015). "The Sustainability Imperative: New Insights on Consumer Expectations." https://www.nielsen.com/us/en/insights/report/2015/the-sustainability-imperative/*
18. *Patagonia. "Worn Wear: Reuse and Repair." https://www.patagonia.com/our-footprint/worn-wear/*

19. *Product-as-a-Service: A New Business Model for Sustainability. BCG. https://www.bcg.com/publications/2019/product-as-a-service*
20. *Rizos, V., et al. (2016). "The Circular Economy: Barriers and Opportunities for Small and Medium-Sized Enterprises." Centre for European Policy Studies. https://www.ceps.eu/ceps-publications/circular-economy-barriers-and-opportunities-small-and-medium-sized-enterprises*
21. *The Circular Economy: A New Agenda for Sustainable Development. OECD. https://www.oecd.org/environment/circular-economy.htm*
22. *Thomas, S. (2024). "Design for a Circular Economy." TEDx Berlin. https://www.ted.com/tedx*
23. *Tukker, A. (2015). "Product services for a resource-efficient and circular economy." Journal of Cleaner Production, 97, 76-91.*
24. *United Nations Environment Programme (UNEP). (2020). "Circular Economy: Policy and Regulation." https://www.unep.org/resources/report/circular-economy-policy-and-regulation*
25. *World Economic Forum. (2021). "Accelerating the Transition to a Circular Economy." [https://www.weforum.org/reports/acceler*

Chapter 11: Role of International Organizations & Political Leaders

International agreements are crucial in shaping global environmental policies and fostering international cooperation to tackle urgent environmental challenges. Key agreements like the Paris Agreement set global climate targets and frameworks for action, influencing policy changes, driving collaboration, and guiding national and regional efforts.

The Paris Agreement, adopted in 2015, aims to limit global warming to well below 2°C, with a goal of 1.5°C, by establishing frameworks such as Nationally Determined Contributions (NDCs) and financial support mechanisms for developing countries. Influential international organizations, including the United Nations (UN), European Union (EU), Organization of Islamic Cooperation (OIC), BRICS, and NATO, play essential roles in implementing and supporting these agreements. They help coordinate efforts, provide financial and technological support, and address environmental impacts through policy integration and monitoring.

Political leadership is integral to advancing sustainable development agendas. Leaders influence public opinion, set policy priorities, build partnerships, address social inequalities, and ensure accountability. Effective leadership can mobilize resources, inspire collective action, and implement strategies that balance economic growth with environmental protection. By leveraging their authority, political leaders drive progress towards a sustainable and equitable future.

The chapter also explores the challenges of international cooperation, including financial and technological disparities, divergent priorities, and governance issues. Addressing these challenges requires enhanced global agreements, technology transfer, inclusive dialogue, and increased financial support to achieve environmental sustainability goals.

11.1 Influence of International Agreements on Global Environmental Policies

International agreements play a pivotal role in shaping global environmental policies and fostering cooperation among nations to address pressing environmental challenges. Agreements such as the Paris Agreement have been instrumental in setting global targets for climate action and environmental sustainability. The influence of such agreements extends beyond mere guidelines; they drive policy changes, foster international collaboration, and set the framework for national and regional actions. To achieve these goals, influential international organizations—including the United Nations (UN), the Organization of Islamic Cooperation (OIC), the European Union (EU), BRICS, NATO, and other regional groups—must actively engage and ensure effective implementation of these agreements. This collective effort is crucial for creating a sustainable future for future generations.

11.1.1 Paris Agreement: A Case Study

The Paris Agreement, adopted in December 2015 at the 21st Conference of the Parties (COP21) to the United Nations Framework Convention on Climate Change (UNFCCC), represents a landmark achievement in international climate policy. Its primary objective is to limit global warming to well below 2°C above pre-industrial levels, with an aspirational goal of limiting the increase to 1.5°C. The agreement establishes a global framework for climate action that includes:

- **Nationally Determined Contributions (NDCs):** Each participating country submits its own climate action plan, known as an NDC, outlining its goals and measures for reducing greenhouse gas emissions and adapting to climate impacts.

- **Global Stocktake:** Every five years, a global stocktake is conducted to assess progress toward the agreement's goals and encourage countries to increase their ambition in their NDCs.
- **Financial Support:** Developed countries are expected to provide financial assistance to developing nations to support their climate actions and adaptation efforts, particularly those that are most vulnerable to climate impacts.
- **Long-Term Goals:** The agreement aims to achieve climate neutrality (net-zero emissions) in the second half of the century and to enhance global efforts to mitigate and adapt to climate change.

11.1.2 Role of International Organizations

- **United Nations (UN):** The UN plays a central role in coordinating international environmental efforts. The UNFCCC, as the primary body overseeing the Paris Agreement, is responsible for facilitating negotiations, monitoring progress, and providing technical support to countries. Other UN agencies, such as the United Nations Environment Programme (UNEP) and the United Nations

Development Programme (UNDP), contribute to environmental protection through research, policy advice, and capacity building.

- **European Union (EU):** The EU has been a strong proponent of climate action and environmental sustainability. It has set ambitious climate targets, such as reducing greenhouse gas emissions by at least 55% by 2030 compared to 1990 levels and achieving climate neutrality by 2050. The EU also promotes environmental policies through its Green Deal and Climate Law, which aim to integrate environmental considerations into all areas of policy and economic activity.
- **Organization of Islamic Cooperation (OIC):** The OIC, representing 57 member states, has been working to address environmental issues within the context of its member countries. The OIC's initiatives include promoting sustainable development, addressing climate change impacts, and fostering cooperation on environmental challenges. The organization has adopted various resolutions and action plans to enhance environmental protection and sustainability among its member states.

- **BRICS:** The BRICS group (Brazil, Russia, India, China, and South Africa) represents major emerging economies with significant environmental impacts. The BRICS countries collaborate on environmental issues through joint initiatives and dialogues. They have committed to addressing climate change and promoting sustainable development in their respective countries and beyond. Their collective efforts contribute to global climate action and environmental policy development.
- **NATO:** While NATO is primarily a military alliance, it has increasingly recognized the links between environmental security and global stability. NATO's policies address the impacts of climate change on security and defence, including the potential for resource conflicts and the need for resilience in military operations. The organization's environmental initiatives focus on reducing the environmental footprint of its activities and promoting climate resilience in member countries.

11.1.3 Regional Groups & Other International Bodies

Various regional organizations, such as the African Union (AU) and the Association of Southeast Asian Nations (ASEAN), also play a role in addressing environmental challenges. These groups often coordinate regional policies, share best practices, and collaborate on cross-border environmental issues. Their efforts complement global agreements and contribute to localized implementation strategies.

11.1.4 Implementation & Impact

- **Policy Integration:** The influence of international agreements on global environmental policies is evident in the integration of environmental considerations into national policies and strategies. Countries that are party to the Paris Agreement are required to align their domestic policies with their NDCs, leading to the development of climate action plans, green regulations, and sustainable development strategies. This integration helps ensure that environmental goals are reflected in economic and social policies.
- **Financial & Technological Support:** International agreements often include provisions for financial and technological support to help

countries, particularly those with limited resources, implement their environmental commitments. Mechanisms such as the Green Climate Fund (GCF) and the Global Environment Facility (GEF) provide funding for climate projects and initiatives. Additionally, technology transfer and capacity-building programs support the adoption of sustainable practices and innovations.

- **Monitoring & Accountability:** Effective implementation of international agreements relies on robust monitoring and accountability mechanisms. Regular reporting, transparency requirements, and independent assessments help track progress and ensure that commitments are met. The global stocktake process under the Paris Agreement, for example, provides a platform for assessing collective progress and encouraging increased ambition.
- **Challenges in Enforcement:** Despite the framework provided by international agreements, enforcement and compliance remain challenging. Many countries struggle to meet their targets due to economic, political, or technical constraints. Ensuring that all parties adhere to their commitments requires on-going international cooperation, support, and pressure from the global community.

The influence of international agreements on global environmental policies is profound and multifaceted. Agreements like the Paris Agreement set ambitious goals and create frameworks for climate action, but their successful implementation depends on the active engagement of international organizations and the effective coordination of global efforts.

Organizations such as the UN, EU, OIC, BRICS, NATO, and regional groups play crucial roles in shaping and enforcing environmental policies. Addressing the challenges of infrastructure, financial barriers, regulatory constraints, and behavioural change is essential for scaling up the impact of these agreements. By leveraging the opportunities for innovation, economic benefits, consumer demand, and policy support, the global community can advance towards a more sustainable and resilient future for all.

11.2 Role of Political Leadership in Driving Sustainable Development Agendas

Sustainable development is a multifaceted and dynamic process aimed at improving the quality of life for current and future generations while balancing economic growth, social inclusion, and environmental protection. Political leadership plays a crucial role in steering this complex agenda, leveraging their influence over public opinion and policy to drive progress. Leaders have the unique ability to shape and promote sustainable development initiatives by mobilizing resources, inspiring collective action, and implementing strategic policies that address both immediate needs and long-term goals.

11.2.1 Influencing Public Opinion & Mobilizing Support

One of the primary roles of political leaders in sustainable development is their ability to influence public opinion. Leaders often wield significant sway over their constituents, shaping the narrative around environmental and social issues through speeches, policies, and public engagement. By framing sustainable development as a priority, political leaders can galvanize public support, fostering a collective sense of responsibility and urgency.

For example, former U.S. President Theodore Roosevelt was instrumental in advancing conservation efforts in the early 20th century. His leadership and commitment to environmental preservation not only helped to establish national parks and monuments but also raised awareness about the importance of natural resource management. In contemporary contexts, leaders like Greta Thunberg, though not a traditional political figure, have demonstrated the power of advocacy in driving global attention to climate change, illustrating how leaders can mobilize youth and inspire widespread movements through strategic communication.

11.2.2 Setting Policy Agendas & Prioritizing Sustainable Development

Political leaders also play a pivotal role in setting policy agendas that prioritize sustainable development. Through legislative action and executive orders, leaders can introduce and enforce policies that promote environmental protection, economic equity, and social justice. The alignment of political will with sustainable development goals can result in substantial and impactful changes.

For instance, the adoption of the Paris Agreement in 2015 was a significant milestone in global climate diplomacy, driven by political leaders from around the world. This international accord aimed to limit global temperature rise and mitigate the effects of climate change through nationally determined contributions (NDCs). The agreement's success hinged on the commitment of political leaders to negotiate, ratifies, and implements policies that align with the accord's targets.

In a more localized context, political leadership at the municipal level can drive sustainable development through initiatives such as urban planning, waste management, and renewable energy projects. Leaders who prioritize green infrastructure and sustainable urban development can influence local economies and improve the quality of life for their constituents.

11.2.3 Building & Sustaining Partnerships

Sustainable development often requires collaborative efforts across various sectors, including government, business, and civil society. Political leaders are in a unique position to forge and sustain these partnerships, leveraging their authority and influence to create synergies that advance sustainability goals.

A notable example is the role of political leaders in fostering public-private partnerships (PPPs) for infrastructure development. PPPs can mobilize private sector investment and expertise to support public projects with sustainable outcomes, such as renewable energy installations, green building initiatives, and efficient transportation systems. Leaders who champion such partnerships can drive innovation and accelerate the implementation of sustainable solutions.

Furthermore, political leaders can play a crucial role in facilitating international cooperation on sustainability issues. For example, the United Nations' Sustainable Development Goals (SDGs) provide a global framework for addressing complex challenges such as poverty, inequality, and climate change. Political leaders who actively participate in international forums and agreements can contribute to collective efforts and ensure that their countries are aligned with global sustainability objectives.

11.2.4 Addressing Inequality & Promoting Social Justice

Sustainable development is not only about environmental and economic considerations but also about promoting social equity and justice. Political leaders have a responsibility to address inequalities and ensure that sustainable development efforts benefit all segments of society. This involves creating policies that reduce poverty, improve access to education and healthcare, and empower marginalized communities.

For instance, leaders who prioritize gender equality and women's empowerment contribute to sustainable development by addressing systemic barriers that hinder women's participation in economic and social activities. The empowerment of women can lead to improved community health, educational outcomes, and economic stability, all of which are integral to sustainable development.

Political leaders must also consider the needs of indigenous and rural communities, who often bear the brunt of environmental degradation and social inequities. Leaders who engage with these communities and incorporate their knowledge and perspectives into policy-making can create more inclusive and effective sustainable development strategies.

11.2.5 Ensuring Accountability & Transparency

Effective political leadership in sustainable development also involves ensuring accountability and transparency in the implementation of policies and programs. Leaders must establish mechanisms to monitor progress, evaluate outcomes, and address challenges that arise. Transparency in decision-making processes and the use of public resources helps build trust and ensures that sustainability efforts are conducted with integrity and effectiveness.

For example, leaders who promote open data initiatives and encourage public participation in policy development can enhance accountability and foster a culture of transparency. This approach allows citizens to be informed about sustainability efforts, participate in decision-making, and hold leaders accountable for their commitments.

Political leadership plays a vital role in driving sustainable development agendas by influencing public opinion, setting policy priorities, building partnerships, addressing social inequalities, and ensuring accountability. Leaders who are committed to sustainability can inspire collective action, implement strategic policies, and mobilize resources to address the complex challenges of our time. By leveraging their influence and authority, political leaders can shape a more sustainable and equitable future for all.

11.3 Analysis of International Cooperation & Challenges in Achieving Environmental Sustainability Goals

Achieving environmental sustainability is a global imperative that necessitates cooperation across nations, yet the international landscape is marked by significant disparities between developed, developing, and under-developed countries. This division complicates collective efforts and poses several challenges to reaching sustainability goals. Understanding these challenges requires a nuanced analysis of how varying levels of development impact cooperation, resource allocation, and policy implementation.

11.3.1 Global Divide: Developed, Developing, & Under-Developed Countries

- **Developed Countries:** Developed countries, characterized by advanced economies and high standards of living, have historically contributed the most to environmental degradation due to industrialization and high consumption patterns. These nations often possess the financial resources, technological expertise, and institutional capacity necessary to address environmental issues. However, their commitment to sustainability is frequently influenced by domestic priorities and economic interests.

  For example, the United States and European Union countries have made significant strides in implementing renewable energy technologies and setting ambitious climate targets. Yet, the challenge remains in reconciling these efforts with economic growth, especially in the context of international trade and economic policies. Developed nations are also expected to take on a leadership role in supporting global sustainability initiatives through funding and technology transfer to less developed regions.

- **Developing Countries:** Developing countries are in a transitional phase, experiencing rapid industrialization and economic growth while grappling with environmental challenges. These nations often face the dual pressures of improving economic conditions and addressing environmental sustainability. The disparity in resources and technological capabilities compared to developed countries makes it difficult for them to implement advanced environmental solutions.

  In many cases, developing countries are highly vulnerable to the effects of environmental degradation and climate change, despite contributing less to the problem. For instance, countries in sub-Saharan Africa are experiencing severe impacts from climate change, such as droughts and floods, which exacerbate existing economic and social challenges. These nations often require substantial international support to build resilience and implement effective sustainability measures.

- **Under-Developed Countries:** Under-developed countries, marked by extreme poverty and limited infrastructure, face significant barriers in addressing environmental sustainability. These nations often prioritize immediate developmental needs over long-term environmental goals, leading to unsustainable practices that further hinder their progress. The lack of resources, weak institutional frameworks, and limited access to technology make it challenging for under-developed countries to participate in global sustainability efforts effectively. Additionally, these nations are frequently the most affected by global environmental issues, such as deforestation and biodiversity loss, despite their minimal contributions to the causes.

11.3.2 Challenges Ahead of International Cooperation

- **Divergent Priorities and Interests:** One of the central challenges in international cooperation is the divergence in priorities and interests among countries. Developed nations often focus on reducing emissions and advancing green technologies, while developing and under-developed countries may prioritize economic growth and poverty alleviation. This divergence can lead to conflicts and delays in reaching consensus on global environmental agreements.

  For example, during the negotiations of the Paris Agreement, developed countries were urged to provide financial and technical support to developing nations to help them transition to low-carbon economies. However, disagreements over the scale and nature of this support have caused tensions and hindered progress. The different priorities and capacities of countries contribute to difficulties in formulating and implementing cohesive and equitable sustainability strategies.

- **Financial & Technological Disparities:** The financial and technological gaps between nations exacerbate the challenges of international cooperation. Developed countries have the means to invest in and deploy advanced technologies that can address environmental issues, such as renewable energy systems and pollution control technologies. In contrast, developing and under-developed countries often lack the necessary resources and infrastructure to access or implement these technologies.

  International financial mechanisms, such as the Green Climate Fund (GCF), have been established to provide support to developing countries. However, the effectiveness of these mechanisms can be limited by bureaucratic hurdles, insufficient funding, and challenges in aligning support with local needs. Bridging the financial and technological divide requires not only increased investment but also effective mechanisms for technology transfer and capacity building.

- **Institutional & Governance Issues:** Institutional and governance challenges also play a significant role in hindering international cooperation. Effective environmental governance requires strong institutions, transparent processes, and accountable leadership. However, many developing and under-developed countries face weaknesses in their governance structures, which can impede their ability to implement and manage sustainability initiatives.

  Moreover, international environmental agreements often require coordinated action at multiple levels of government, which can be difficult to achieve in countries with fragmented or weak institutions. Strengthening governance frameworks and fostering collaboration between national and international bodies are crucial for improving the effectiveness of global sustainability efforts.

- **Global Environmental and Economic Policies:** The interplay between global environmental policies and economic systems can create further obstacles to cooperation. Trade policies, economic sanctions, and market dynamics can influence a country's willingness and ability to engage in sustainability efforts. For instance, economic incentives such as subsidies for fossil fuels can undermine efforts to transition to cleaner energy sources.

Additionally, international trade agreements and economic competition can impact countries' environmental policies. Developing and under-developed countries may be reluctant to adopt stringent environmental regulations if they perceive that such measures could disadvantage their industries in global markets. Balancing economic growth with environmental sustainability requires careful policy design and international collaboration.

11.3.3 Pathways to Enhanced International Cooperation

To overcome these challenges and achieve environmental sustainability goals, a multifaceted approach is required:

- **Strengthening Global Agreements:** Enhancing the effectiveness of international environmental agreements through clear targets, equitable resource distribution, and robust monitoring mechanisms can improve cooperation and accountability.

- **Promoting Technology Transfer & Capacity Building:** Facilitating the transfer of green technologies and providing capacity-building support to developing and under-developed countries can help bridge the technological and financial gaps.
- **Fostering Inclusive Dialogue:** Engaging all stakeholders, including governments, businesses, and civil society, in inclusive dialogues can help reconcile differing priorities and develop more effective and equitable solutions.
- **Enhancing Financial Support Mechanisms:** Increasing and streamlining financial support for sustainability initiatives, while ensuring that funding aligns with local needs and conditions, is crucial for empowering countries to take meaningful action.

By addressing these challenges and leveraging collective strengths, the global community can make significant strides towards achieving environmental sustainability and ensuring a more equitable and resilient future for all.

11.4 References

1. *African Union (AU). (2020). "African Union's Agenda 2063 and Environmental Sustainability." https://au.int/en/agenda2063*
2. *Association of Southeast Asian Nations (ASEAN). (2021). "ASEAN Environment Action Plan." [https://asean.org/asean-environment action-plan/](https://asean.org/asean-environment-action-plan/)*
3. *Bernstein, S. (2001). The Green Chain: International Environmental Negotiations and the Formation of Global Governance. Cambridge University Press.*
4. *Bodansky, D. (2016). The Paris Agreement and the New Logic of International Climate Politics. Harvard Law Review, 130(6), 1886-1910.*
5. *BRICS. (2021). "BRICS Joint Statement on Climate Change." https://www.brics2021.gov.br/*
6. *Diamond, J. (2005). Collapse: How Societies Choose to Fail or Succeed. Penguin Books.*
7. *European Commission. (2020). "The European Green Deal." https://ec.europa.eu/info/strategy/priorities-2019-2024/european-green-deal_en*
8. *Garnaut, R. (2008). The Garnaut Climate Change Review. Cambridge University Press.*
9. *Hirsch, A. (2005). Righteous Rebellion: The Role of Civil Society in Sustainable Development. Routledge.*
10. *Keohane, R. O., & Victor, D. G. (2011). The Regime Complex for Climate Change. Perspectives on Politics, 9(1), 7-23.*

11. *NATO. (2020). "Climate Change and Security." https://www.nato.int/cps/en/natolive/topics_156810.htm*
12. *Organization of Islamic Cooperation (OIC). (2020). "OIC Action Plan for Environment and Climate Change." https://www.oic-oci.org/*
13. *Sachs, J. D. (2015). The Age of Sustainable Development. Columbia University Press.*
14. *Sachs, J. D. (2015). The Age of Sustainable Development. Columbia University Press.*
15. *United Nations Development Programme (UNDP). (2020). Human Development Report 2020: The Next Frontier: Human Development and the Anthropocene. UNDP.*
16. *United Nations Environment Programme (UNEP). (2021). "Annual Report." https://www.unep.org/resources/report/unep-annual-report-2021*
17. *United Nations Framework Convention on Climate Change (UNFCCC). (2015). "Paris Agreement." https://unfccc.int/sites/default/files/english_paris_agreement.pdf*
18. *United Nations Framework Convention on Climate Change (UNFCCC). (2020). Annual Report. UNFCCC Secretariat.*

Chapter 12: Critical Analysis & Hope for Sustainable Futures

The quest for a sustainable future presents a complex challenge that necessitates a comprehensive examination of current practices, policies, and global dynamics. Sustainable development, defined as meeting present needs without compromising the future, addresses a broad spectrum of issues including environmental protection, social equity, and economic growth. This chapter provides a critical analysis of sustainable development, offering insights into the current state and exploring potential pathways toward a more hopeful future.

Key insights into sustainable development highlight pressing issues such as environmental degradation, climate change, and socio-economic disparities. Industrial activities and pollution have led to significant ecological imbalances and climate impacts, requiring urgent mitigation and adaptation measures. Additionally, disparities between developed and developing nations create challenges in implementing sustainable practices, emphasizing the need for equitable support.

The chapter underscores the vital role of policy and governance, noting the gaps in implementing international agreements such as the Paris Accord and the barriers posed by financial and technological inequities. Technological innovations and multi-stakeholder collaborations offer hope, with advancements in renewable energy and successful international partnerships demonstrating progress.

Furthermore, the chapter emphasizes the importance of advancing environmental education and lifelong learning. Integrating environmental topics into curricula, promoting interdisciplinary higher education, and enhancing community-based learning are essential for fostering a well-informed and proactive society.

To drive meaningful change, the chapter calls for collective action from individuals, communities, and policymakers. By adopting sustainable practices, supporting environmental advocacy, and developing equitable policies, society can work towards a resilient and sustainable future.

12.1 Summary of Key Insights & Findings

The quest for a sustainable future is a formidable challenge that demands an in-depth analysis of current practices, policies, and global dynamics. Sustainable development, defined as meeting the needs of the present without compromising the ability of future generations to meet their own needs, encompasses a broad array of issues including environmental protection, social equity, and economic growth. This critical analysis delves into the current state of sustainable development, examining key insights and findings while exploring potential pathways for a more hopeful and sustainable future.

12.1.1 Key Insights into Sustainable Development

- **Environmental Degradation & Climate Change:** Environmental degradation, driven by industrial activities, deforestation, and pollution, poses a significant threat to the planet's ecological balance. Climate change, a major consequence of these activities, is leading to more frequent and severe weather events, rising sea levels, and disruptions to ecosystems. According to the Intergovernmental Panel on Climate Change (IPCC), the impacts of climate change are already being felt across the globe, affecting agriculture, water resources, and biodiversity. The urgent need to mitigate greenhouse gas emissions and adapt to changing climate conditions is a central theme in sustainability discussions.

- **Socio-Economic Disparities:** The global divide between developed, developing, and under-developed countries create significant challenges for achieving sustainability. Developed nations, with their advanced economies and technological capabilities, have historically contributed the most to environmental degradation but also have the resources to address it. Conversely, developing and under-developed countries, often with limited resources and institutional capacity, face considerable hurdles in implementing sustainable practices. These disparities lead to unequal burdens of environmental problems and require targeted support to address the needs of less affluent nations.
- **The Role of Policy & Governance:** Effective governance and policy frameworks are crucial for advancing sustainability. Policies must integrate environmental, social, and economic dimensions to be effective. International agreements, such as the Paris Agreement, represent significant efforts to create a unified approach to climate change, but implementation remains inconsistent. National and local policies are also critical, as they translate global commitments into actionable measures. Effective governance involves not only setting

ambitious targets but also ensuring transparency, accountability, and public engagement.

- **Technological Innovations & Solutions:** Technological advancements play a vital role in promoting sustainability. Innovations in renewable energy, waste management, and resource efficiency offer promising solutions to many environmental challenges. Solar and wind energy, electric vehicles, and smart grid technologies are examples of how technology can drive the transition to a low-carbon economy. However, the deployment of these technologies often requires substantial investment and supportive policies, particularly in developing and under-developed regions.
- **Importance of Multi-Stakeholder Collaboration:** Addressing sustainability challenges requires collaboration among governments, businesses, non-governmental organizations (NGOs), and communities. Multi-stakeholder partnerships can enhance the effectiveness of sustainability initiatives by leveraging diverse expertise and resources. For example, public-private partnerships can accelerate the development and implementation of green technologies,

while community-based organizations can drive local conservation and education efforts.

12.1.2 Critical Analysis of Current Challenges

- **Implementation Gaps in International Agreements:** Despite the establishment of international frameworks and agreements, there are significant gaps in implementation. The Paris Agreement, while a landmark achievement in climate diplomacy, has faced challenges in translating commitments into concrete actions. Variability in national priorities, financial constraints, and lack of enforcement mechanisms have hindered progress. Ensuring that countries meet their climate targets requires robust monitoring systems, regular reporting, and mechanisms for accountability.
- **Financial & Technological Inequities:** The disparity in financial and technological resources between developed and less developed countries creates barriers to achieving sustainability goals. Developed nations have the means to invest in advanced technologies and infrastructure, while developing and under-developed countries often struggle with basic needs and lack access to cutting-edge solutions.

Bridging this gap requires increased international financial support, technology transfer, and capacity-building initiatives to empower all countries to participate in and benefit from sustainability efforts.

- **Resistance to Change:** Resistance to change, whether from entrenched economic interests, political opposition, or cultural factors, can impede progress toward sustainability. Fossil fuel industries, for example, often resist regulations that threaten their profits, while political leaders may prioritize short-term economic gains over long-term environmental benefits. Addressing this resistance requires strategic communication, stakeholder engagement, and incentives that align economic and environmental objectives.
- **Institutional Weaknesses:** Weak institutional frameworks in many countries hinder the effective implementation of sustainability policies. Corruption, lack of transparency, and insufficient capacity can undermine efforts to address environmental and social issues. Strengthening institutions, improving governance, and fostering public trust are essential for creating a supportive environment for sustainable development.

12.1.3 Hope for a Sustainable Future

- **Growing Awareness & Advocacy:** There is a growing global awareness and advocacy for sustainability, driven by civil society, youth movements, and environmental organizations. Grassroots movements, such as Fridays for Future and Extinction Rebellion, have galvanized public support and pressured governments and businesses to take more ambitious actions. This increased awareness and activism can drive positive change by highlighting issues, influencing policies, and promoting sustainable practices.
- **Advances in Sustainable Technology:** Technological innovations continue to offer new solutions for sustainability challenges. Advances in renewable energy, energy storage, and circular economy practices provide opportunities for reducing environmental impacts and enhancing resource efficiency. On-going research and development efforts are likely to yield even more effective and affordable solutions in the future.
- **Collaborative Initiatives:** Collaborative initiatives, including international agreements, multi-stakeholder partnerships, and local community projects, offer hope for achieving sustainability goals.

Successful examples of collaboration, such as the reforestation efforts in Costa Rica or the transition to renewable energy in Denmark, demonstrate the potential of working together to address environmental issues and create sustainable solutions.

- **Policy & Regulatory Developments:** There is a growing trend toward stronger environmental policies and regulations at both national and international levels. Governments are increasingly incorporating sustainability into their agendas, setting ambitious targets, and implementing policies to reduce emissions and protect natural resources. Continued progress in policy development, combined with effective enforcement, can support the transition to a more sustainable future.

12.2 Recommendations for Advancing Environmental Education & Lifelong Learning

Environmental education and lifelong learning are critical components in fostering a sustainable future. As global environmental challenges escalate, it becomes imperative to equip individuals with the knowledge, skills, and attitudes necessary to address these issues effectively. Advancing environmental education and lifelong learning can empower people to make informed decisions, adopt sustainable practices, and contribute to environmental stewardship. This analysis offers recommendations for enhancing environmental education and lifelong learning to build a more sustainable and informed society.

12.2.1 Integrating Environmental Education into Formal Curricula

- **Curriculum Development:** Integrating environmental education into formal curricula from an early age is essential for fostering environmental literacy. Curricula should include topics related to ecology, climate change, sustainability, and conservation across various subjects. For instance, science classes can cover the principles of ecosystems and climate science, while social studies can explore the impact of environmental policies and historical case studies of environmental movements. By embedding environmental themes into core subjects, students gain a holistic understanding of the interconnectedness of human and natural systems.
- **Teacher Training & Resources:** Effective environmental education relies on well-trained educators who are knowledgeable about

environmental issues and teaching methodologies. Professional development programs for teachers should include training on environmental topics, pedagogical strategies for teaching sustainability, and access to up-to-date resources. Providing teachers with lesson plans, educational materials, and access to environmental experts can enhance their ability to deliver engaging and informative lessons on environmental issues.

- **Hands-On Learning & Experiential Education:** Experiential education, including field trips, outdoor learning experiences, and hands-on projects, can significantly enhance students' understanding of environmental concepts. Activities such as nature walks, community clean-ups, and school gardens allow students to apply classroom knowledge in real-world contexts, fostering a deeper connection to environmental issues. Schools should prioritize opportunities for students to engage with their local environment and participate in environmental stewardship activities.

12.2.2 Promoting Environmental Literacy through Higher Education

- **Interdisciplinary Programs:** Higher education institutions should develop interdisciplinary programs that integrate environmental studies with fields such as economics, engineering, and social sciences. These programs can prepare students to address complex environmental challenges by combining technical knowledge with an understanding of economic, social, and policy dimensions. For example, a degree program in environmental engineering might include coursework in sustainable business practices and environmental policy, equipping graduates with a broad skill set for addressing sustainability issues.
- **Research & Innovation:** Encouraging research and innovation in environmental science and sustainability is crucial for advancing knowledge and developing new solutions. Universities and research institutions should support projects that address pressing environmental challenges, such as climate change, resource management, and pollution. Collaboration between academic institutions, government agencies, and private sector partners can

drive innovation and facilitate the application of research findings to real-world problems.

- **Community Engagement & Partnerships:** Higher education institutions should engage with local communities and organizations to promote environmental literacy and address local environmental issues. Partnerships with community groups, businesses, and government agencies can provide students with practical experience and opportunities to contribute to local sustainability efforts. Initiatives such as community-based research projects, environmental internships, and public lectures can bridge the gap between academic knowledge and community needs.

12.2.3 Enhancing Lifelong Learning Opportunities

- **Public Awareness Campaigns:** Public awareness campaigns can play a significant role in promoting environmental education and encouraging lifelong learning. These campaigns should aim to increase public understanding of environmental issues, highlight the importance of sustainability, and provide information on how individuals can contribute to environmental protection. Media outlets, social media platforms, and community events can be used to disseminate information and engage the public in environmental learning.

- **Online Courses & Resources:** Online platforms offer flexible and accessible opportunities for lifelong learning in environmental education. MOOCs (Massive Open Online Courses), webinars, and online resources can provide individuals with the knowledge and skills needed to address environmental challenges. Courses on topics such as climate change, sustainable living, and conservation can be designed for diverse audiences, including professionals, students, and the general public.
- **Community-Based Learning:** Community-based learning initiatives can foster lifelong environmental education by involving individuals in local sustainability efforts and providing opportunities for hands-on experience. Programs such as community workshops, citizen science projects, and environmental volunteer opportunities can engage people of all ages in learning about and contributing to environmental protection. These initiatives can also promote a sense of community and collective responsibility for environmental stewardship.

- **Integration with Professional Development:** Incorporating environmental education into professional development programs can help individuals apply sustainability principles in their careers. Employers and professional organizations can offer training and resources on sustainable practices, environmental regulations, and corporate social responsibility. By integrating environmental considerations into professional development, individuals can contribute to sustainability efforts within their industries and workplaces.

12.2.4 Addressing Challenges & Ensuring Equity

- **Overcoming Barriers to Access:** Ensuring equitable access to environmental education is essential for fostering widespread engagement and participation. Efforts should be made to address barriers such as socioeconomic disparities, geographic isolation, and lack of resources. Scholarships, community outreach programs, and online learning opportunities can help ensure that individuals from diverse backgrounds have access to environmental education and lifelong learning resources.

- **Evaluating & Adapting Programs:** Regular evaluation and adaptation of environmental education programs are necessary to ensure their effectiveness and relevance. Feedback from students, educators, and community members can provide valuable insights into program strengths and areas for improvement. Continuous assessment and adaptation can help address emerging environmental issues and incorporate new educational approaches and technologies.
- **Building Collaborative Networks:** Collaboration between educational institutions, government agencies, non-profit organizations, and businesses can enhance the effectiveness of environmental education initiatives. By working together, these stakeholders can share resources, expertise, and best practices to develop and implement effective environmental education programs. Collaborative networks can also help identify and address gaps in environmental education and promote innovative approaches to lifelong learning.

12.3 Call to Action for Individuals, Communities, & Policymakers: Prioritizing Sustainability in Decision-Making

The urgency of addressing environmental issues cannot be overstated. Climate change, resource depletion, and ecological degradation threaten the stability of our planet and the well-being of future generations. To confront these challenges effectively, it is essential that every segment of society, individuals, communities, and policymakers, embraces sustainability in their decision-making processes. Only through collective and concerted action can we safeguard our environment and ensure a sustainable future.

12.3.1 Role of Individuals in Sustainability

- **Adopt Sustainable Practices:** Individuals play a critical role in driving sustainability through their everyday choices and behaviours. Simple actions, such as reducing energy consumption, conserving water, minimizing waste, and choosing sustainable products, can collectively lead to significant environmental benefits. For example, using energy-efficient appliances, reducing single-use plastics, and supporting sustainable agriculture can contribute to lowering one's carbon footprint and reducing environmental impact.
- **Educate & Advocate:** Personal commitment to sustainability extends beyond individual actions to include education and advocacy.

Individuals can educate themselves and others about environmental issues and the importance of sustainable practices. By raising awareness through social media, community events, and conversations with peers, individuals can influence others and foster a culture of environmental responsibility. Advocacy efforts can also involve participating in local environmental campaigns, supporting green policies, and engaging with organizations dedicated to sustainability.

- **Support Sustainable Businesses:** Consumer choices have a powerful impact on the market. By supporting businesses that prioritize sustainability, individuals can drive demand for eco-friendly products and services. This includes purchasing from companies with sustainable supply chains, investing in green technologies, and favouring businesses with transparent environmental practices. Consumer pressure can incentivize more companies to adopt sustainable practices and contribute to broader environmental goals.

12.3.2 Role of Communities in Sustainability

- **Foster Local Initiatives:** Communities are well-positioned to implement and support local sustainability initiatives. Community-based projects, such as urban gardens, recycling programs, and local clean-up events, can significantly impact environmental quality and community well-being. By organizing and participating in these initiatives, communities can address local environmental issues, promote sustainable living practices, and enhance community cohesion.
- **Promote Education & Awareness:** Local educational institutions and community organizations play a vital role in promoting environmental literacy. Schools, libraries, and community centres can offer workshops, lectures, and programs focused on sustainability and environmental issues. By integrating environmental education into community activities, local organizations can raise awareness and empower residents to make informed decisions about their environmental impact.
- **Build Partnerships & Collaborations:** Effective sustainability efforts often require collaboration among various stakeholders,

including local governments, businesses, and non-profit organizations. Communities can facilitate partnerships to pool resources, share expertise, and coordinate actions towards common sustainability goals. For instance, collaborations between local governments and businesses can lead to the development of green infrastructure projects, while partnerships with non-profits can support conservation and environmental education programs.

- **Advocate for Local Policy Changes:** Communities can also influence local policy decisions by advocating for sustainability-focused regulations and initiatives. Engaging with local government officials, participating in public forums, and supporting policies that promote environmental protection can drive systemic changes. Community advocacy can result in policies that encourage sustainable development, improve waste management, and enhance environmental conservation efforts.

12.3.3 Role of Policymakers in Sustainability

- **Develop & Implement Sustainable Policies:** Policymakers have the authority to shape environmental outcomes through legislation and policy-making. Developing and implementing policies that promote sustainability is crucial for addressing environmental challenges at a systemic level. This includes enacting regulations that limit greenhouse gas emissions, protect natural resources, and promote renewable energy. Effective policies can drive significant progress towards sustainability goals and set a standard for environmental stewardship.

- **Support Research & Innovation:** Policymakers can support sustainability by funding research and fostering innovation in green technologies and practices. Investments in research can lead to the development of new solutions for environmental problems, such as advanced renewable energy technologies, sustainable agriculture practices, and pollution control methods. By prioritizing research and development, policymakers can facilitate the transition to a more sustainable economy and enhance environmental resilience.
- **Facilitate Collaboration & Engagement:** Policymakers should facilitate collaboration among stakeholders, including government agencies, businesses, and civil society organizations. Collaborative approaches can lead to more comprehensive and effective

sustainability strategies. For example, multi-stakeholder partnerships can help address complex environmental issues, such as climate change and resource management, by integrating diverse perspectives and expertise.

- **Ensure Equity & Inclusivity:** Sustainability efforts must be inclusive and equitable, addressing the needs of all segments of society. Policymakers should ensure that environmental policies consider social equity and do not disproportionately impact marginalized communities. This involves designing policies that provide support to vulnerable populations, promote environmental justice, and ensure that all individuals have access to clean air, water, and natural resources.
- **Monitor & Report Progress:** Effective environmental policies require robust monitoring and reporting mechanisms. Policymakers should establish systems to track progress towards sustainability goals, evaluate the effectiveness of policies, and ensure transparency. Regular reporting can provide insights into policy impacts, identify areas for improvement, and maintain public accountability.

- **Collective Call to Action:** Addressing environmental challenges requires a unified effort from individuals, communities, and policymakers. Each group has a unique role to play, and their collective actions can drive meaningful change. Individuals must make sustainable choices and advocate for environmental responsibility. Communities should implement local initiatives, promote education, and build partnerships. Policymakers need to develop and enforce sustainable policies, support research, and ensure equity.

By prioritizing sustainability in decision-making and taking coordinated action, we can create a resilient and sustainable future. The path forward involves recognizing our shared responsibility and working together to protect the environment for current and future generations. The time for action is now, and every effort counts in the pursuit of a sustainable and equitable world.

12.4 References

1. *Intergovernmental Panel on Climate Change (IPCC). (2021). Climate Change 2021: The Physical Science Basis. Cambridge University Press.*
2. *Intergovernmental Panel on Climate Change (IPCC). (2021). Sixth Assessment Report: Climate Change 2021: The Physical Science Basis. Cambridge University Press.*
3. *McKeown, R., & Hopkins, C. (2003). Environmental Education: A Practical Guide for Educators. Routledge.*
4. *Sachs, J. D. (2015). The Age of Sustainable Development. Columbia University Press.*
5. *Sachs, J. D. (2015). The Age of Sustainable Development. Columbia University Press.*
6. *Tilbury, D. (2011). Education for Sustainable Development: An Expert Review of Processes and Learning. UNESCO.*
7. *UNESCO. (2014). Education for Sustainable Development Goals: Learning Objectives. United Nations Educational, Scientific and Cultural Organization.*
8. *United Nations Development Programme (UNDP). (2021). Human Development Report 2021: Uncertain Times, Unsettled Lives: Shaping our Future in a Transforming World. UNDP.*
9. *United Nations Environment Programme (UNEP). (2020). Emissions Gap Report 2020. UNEP.*
10. *United Nations Environment Programme (UNEP). (2020). Emissions Gap Report 2020. UNEP.*
11. *United Nations Framework Convention on Climate Change (UNFCCC). (2021). The Paris Agreement: A Historic Milestone for Climate Action. UNFCCC Secretariat.*

12. *Wals, A. E. J., & Corcoran, P. B. (2012). Learning for Sustainability in Times of Accelerating Change. Wageningen Academic Publishers.*
13. *Wals, A. E. J., & Corcoran, P. B. (2012). Learning for Sustainability in Times of Accelerating Change. Wageningen Academic Publishers.*
14. *World Resources Institute (WRI). (2020). Creating a Sustainable Future: Technological Innovations and Solutions. WRI.*

Printed by Books on Demand GmbH, Norderstedt / Germany